German A. Golitsyn
Vladimir M. Petrov

Information and Creation

Integrating the "Two Cultures"

Birkhäuser Verlag
Basel · Boston · Berlin

Authors' Address:

Russian Institute of Art Studies
Kozitsky pereulok 5
Moscow 103009, Russia

e-mail: rubin@riar.msc.su

A CIP catalogue record for this book is available from the Library of Congress, Washington D.C., USA

Deutsche Bibliothek Cataloging-in-Publication Data

Golicyn, German A.:
Information and creation : integrating the "Two cultures" /
German A. Golitsyn ; Vladimir M. Petrov. - Basel ; Boston ;
Berlin : Birkhäuser, 1995

NE: Petrov, Vladimir M.:

Softcover reprint of the hardcover 1st 1995
Printed on acid-free paper produced from chlorine-free pulp ∞
Cover design: Markus Etterich, Basel

ISBN 978-3-0348-7474-8 ISBN 978-3-0348-7472-4 (eBook)
DOI 10.1007/978-3-0348-7472-4

9 8 7 6 5 4 3 2 1

Contents

Preface

This book is devoted to one of the central problems of contemporary thinking, for which C.P. Snow in 1959 coined the phrase of the «Two Cultures». In this concept, human endeavour is directed on one side to the (forward-looking) sciences (mathematics, physics, chemistry, biology, etc.) and on the other side to the (backward-looking) humanities (including psychology, linguistics, sociology, etc.). In this dichotomy Snow saw no possibility of unification.

On the other hand the urge towards self-consistency and harmony in the mental and spiritual lives of both man and society as a whole is clearly one of the major forces of creativity, both scientific and artistic. This force aims at the unification of the «Two Cultures» in order to build an integrated self-consistent system for our intellectual life. Some attempts in this direction have been made before, and will be described in this book.

It is our aim to contribute to the achievement of an integrated mental life on the basis of information theory. In order to construct our model, we examine the laws of information theory, leading us to the deduction of the main laws inherent in both «cultures». Thus, we consider the evolution of both non-living and living matter, human behaviour, the phenomenon of language, the sphere of aesthetics, etc.

We hope that our work will be useful both for researchers (who are trying to derive different integral theories) and for various other «consumers» of scientific knowledge (meaning broad circles of intellectuals).

This book is the result of more than twenty years of investigations of the P.K. Anokhin Institute for Normal Physiology and the Russian Institute of Art Studies (Moscow). The work became possible thanks to the support of the Laboratory of Cognitive Processes of the Moscow Institute of Radio, Electronics and Automation. We are also grateful to our colleagues of the Psychology of Creativity Section of the Moscow Psychological Society headed by Prof. Ya. Ponomaryov — for the fruitful discussions which have accompanied our work during all these years, and to Cathy Beyers — for proofreading and style editing.

The Introduction, Chapters 1, 2 and paragraphs 5.1–5.4 (Chapter 5) were made by German A. Golitsyn; Chapters 3, 4, paragraphs 5.5–5.6 and the Concluding remarks were written by Vladimir M. Petrov.

Moscow, December 1992

German A. Golitsyn
Vladimir M. Petrov

Introduction

One of the main dangers of modern development is the strikingly low effectiveness of human knowledge. Great expenditures of human resources — work, time, energy, etc. give an insignificant result. Now we can control a space ship at the distance of millions of kilometers from the Earth; but we cannot control our own behaviour. And the ability to control is the main criterion for the effectiveness of knowledge.

Some people might say that it is better this way. Because if the technique of control got in the hands of beasts, people would feel bad. But we think it isn't all that bad. Not too many people with devilish inclinations do evil for evil's sake. Most evil is done by people who want to do good. But either they don't know what true good is, or they don't know the true path to it. In other words — they cannot control. We cannot overcome our own weaknesses — laziness, envy, jealousy; we cannot turn our children into genuine people; we cannot beat drugs and terrorism; we cannot tame human greed which could bring a global catastrophe upon us. A man's power grows much faster than his ability for self-control.

In all these cases we try to control, but do it badly. And we so wish we could do it well. It seems to us that some of the experience of the exact sciences could help us in this situation.

Let us begin from the question: what is an explanation in the exact sciences? What is the superior form of scientific explanation — scientific theory? What peculiarities make the theories in exact sciences into «genuine» theories? Some disciplines in humanities claim to be a «theory», but a physicist would rather reject such pretensions. Why?

Every «genuine» scientific theory has two levels. The first one connects facts to each other by means of laws. But there must be a second level where the laws themselves are connected by logical relations. Without this second logical level the theory is not a theory as such — it is a totality of empirical knowledge only. Let us take psychology as example. There are many empirical regularities and facts, and many of them have a mathematical form (the Weber-Fechner law, the Yerx-Dodson law, etc). But the theory — from the physicist's point of view — is absent. The reason for this is that these regularities are isolated from one another, they don't form a logical system, and aren't deduced from one another. Thus in psychology we have knowledge, but no understanding. We don't understand why this law is like this but not like that. In the words of A. Einstein: we cannot «explain empirical regularity as logical necessity».

Let us compare these two facts — the ineffectiveness of human knowledge and the absence of «genuine» theory — and we have to admit that the creation of a theoretical basis of humanity is a problem of paramount importance.

Furthermore, the logical structure of a theory is not invariable, it evolves towards an «ideal», «optimal» form. A tendency of this evolution was already formulated by Newton: to explain the maximum of facts by using the minimum of initial assumptions (postulates, principles). Roughly speaking we can define the effectiveness of the scientific theory E as a relation:

$$E = \frac{\text{Number of facts explained}}{\text{Number of assumptions used}} \tag{1}$$

In the process of the theory's evolution the number of facts explained must increase and the number of initial assumptions must decrease. We can call this tendency «*centralization*».

This tendency towards centralization obviously is not accidental. It is a general regularity inherent to the development of many types of systems: biological, technical, social, etc. Some examples:

Originally, telephone nets connected subscribers to each other directly. But with the growth of the number of subscribers n the number of connections N grew approximately as n^2:

$$N = n(n-1)/2. \tag{2a}$$

The system became complicated very fast. Simplification has been achieved by the introduction of central telephone exchanges. Now each subscriber was connected with the exchange and only by means of it — with the other subscribers. The number of connections had been sharply diminished:

$$N = n. \tag{2b}$$

The system of commerce also evolved towards centralization. At first one product was exchanged for another one directly. But with the growth of the commodity numbers, one of the products stood out among others (usually it was gold) and became the «centre of exchange». Each product was exchanged for gold and then gold — for any other product. The system of exchange became simplified.

One can find similar examples in many other areas of life: the evolution of the nervous system from diffused to centralized, the evolution of political relations from the feudal to the centralized states, and so on. This aspiration towards centralization also moves human creative activity. It is Abraham who searches a single God controlling the multifarious natural and spiritual phenomena. It is Newton who searches a single Law of Universal Gravitation which explains the fall of apples, the motion of the Moon and the planets, the tides and many other facts. One can suppose that such regularity is also inherent to the development of scientific theory.

At the end of the evolution of a scientific theory usually only one central postulate (principle) remains. As a rule it has the form of the optimality principle (or variational principle) — like the principle of the least action in mechanics or the principle of the entropy maximum in thermodynamics and so on.

This is not a coincidence. We know today that the optimality principle is the most compact, capacious, powerful form to express the laws of Nature (Polak,

1960; Lanzcos, 1962; Rosen, 1967). It seems to be very natural to look for the laws of a new area of science (such as psychology, sociology, aesthetics, etc.) — to look for these laws in the form of the optimality principle. Such a principle is connected with particular regularities and phenomena by means of logical and mathematical connections. And these regularities can be deduced from a principle by means of logical and mathematical methods.

We can extract one more important lesson from the experience of the exact science: usually the «principles» of the behaviour of any system are formulated before the «mechanisms» which realize these principles. Newton had formulated the Law of Universal Gravitation before the mechanism of gravitation became understood. Is this mechanism just a «whirlwind» (Descartes), or some «particles» (Lessage), a field or something else? «Hypotheses non figno» was Newton's answer to these questions. So far we don't know what a mechanism of gravitation is, but this does not prevent us from solving various practical problems on the basis of this «principle» — the Law of Gravitation.

In living systems we come across many various mechanisms. Most of them we cannot understand until we know what they are doing, what the «principles» — purposes, functions, and laws — are that are realized by these mechanisms. We could not understand the construction and the work of the heart without knowing the principle of the construction and the work of a pump. Usually we cannot even distinguish what is main and what is secondary in a mechanism, if we don't know what this mechanism does.

The secondary role of mechanisms (as compared to principles) in living systems is obvious from the fact that the same function can be accomplished by quite different mechanisms. It is the reason why the principles of human behaviour will be the main object of our attention. Investigations of mechanisms we consider a task for the second stage.

In biology and psychology some optimality principles were formulated such as «the principle of the least effort» (Zipf, 1949), «the principle of optimal construction» (Rashevsky, 1960), etc. As a rule all these principles are connected with the idea of the economy of any «resource» R (energy, time, space, materials, number of operations, etc):

$$R(y) = \min_{y}, \tag{3}$$

where y describes parameters, traits or responses of an organism. In general, case y is a vector, i.e. has many components. Condition (3) means that the organism selects the values of y (among the set of possible ones) that provide an optimal value of the resource expenses R. This selection can result from the evolution of the species or from individual learning. Usually energy is considered as such a resource R.

The principle of the economy of energy (Rashevsky, 1960; Rosen, 1967; Khanin *et al.*, 1978) is now very popular in biology. It can explain numerous facts and empirical regularities. But on the other hand many facts cannot be explained on the basis of this principle. The general reason is that energy for the living being is the means but not the aim. Economy of energy is permissible only under the following condition: it must not prevent the system from achieving its aim. Thus the question arises: what is the aim in itself?

The aim must be related to the idea of the adaptation of a system to its environment. This adaptation is a necessary condition for the survival of the system in this environment. Any existing system interacts with its environment and has to either adapt to that environment, or be destroyed by it, so that «adaptability» is almost a synonym for «existence». If we had a quantitative measure of adaptation, we could formulate a corresponding optimality principle. It seems that the mutual information between the conditions of the environment and the responses (parameters, traits) of the system is the most suitable measure of adaptation. Thus the concepts of information, adaptation, and the relations between them become the first object of our consideration.

Chapter 1
Information and adaptation

The reasonable man adapts himself to the world; the unreasonable one persists in trying to adapt the world to himself. Therefore all progress depends on the unreasonable man.

Bernard Shaw

Information, as well as matter and energy, is one of the three whales on which the Universe stands. For any level of organization of matter we can write:

$$\text{matter} = \text{energy} + \text{information}.$$

The old idea of «form» belonged to Aristotle, later the concepts of «structure», «organization», etc. acquired an exact quantitative expression and new life with the rise of the concept of information.

Just at the point of intersection of these two concepts — energy and information — we must look for the main laws of life. This is also true for human behaviour and human creativity.

1.1 Mutual information

It is known that mutual information between two variables x and y can be defined (Fano, 1961) through the symmetrical expression:

$$I(x,y) = \log \frac{p(x,y)}{p(x)p(y)} \tag{1.1a}$$

where $p(x)$, $p(y)$ are probabilities of the values x and y; and $p(x,y)$ is the joint probability of the combination of x and y.

Normally we will use a logarithm in base 2, so that the amount of information will be expressed in «bits» — binary units.

Sometimes it is more favourable to use the conditional probabilities $p(x/y) = p(x,y)/p(y)$ and $p(y/x) = p(x,y)/p(x)$.

In this case expression (1.1a) can be rewritten as:

$$I(x,y) = \log \frac{p(y/x)}{p(y)} = \log \frac{p(x/y)}{p(x)} \tag{1.1b}$$

Usually we interpret x as the conditions of the environment («stimuli»); and y as the responses of the system.

Now we must define an average mutual information between variables x and y:

$$\begin{aligned} I(X,Y) &= \sum_x \sum_y p(x,y) \log \frac{p(x,y)}{p(x)p(y)} = \\ &= -\sum_x \sum_y p(x,y) \log p(x) \\ &\quad -\sum_x \sum_y p(x,y) \log p(y) + \sum_x \sum_y p(x,y) \log p(x,y) \\ &= H(X) + H(Y) - H(X,Y) \end{aligned} \tag{1.2a}$$

where $H(X) = -\sum_x \sum_y p(x,y) \log p(x)$ is named the unconditional entropy of stimuli; it can be interpreted also as the «self information», contained in variable x; it is the maximal amount of information which can be received from this variable; $H(Y) = -\sum_x \sum_y p(x,y) \log p(y)$ — entropy of responses; $H(X,Y) = -\sum_x \sum_y p(x,y) \log p(x,y)$ — joint entropy of the system of both variables.

Using conditional probabilities we can also present the average mutual information in two equivalent forms:

$$I(X,Y) = \sum_x p(x) \sum_y p(y/x) \log \frac{p(y/x)}{p(y)} = H(Y) - H(Y/X) \tag{1.2b}$$

$$I(X,Y) = \sum_y p(y) \sum_x p(x/y) \log \frac{p(x/y)}{p(x)} = H(X) - H(X/Y) \tag{1.2c}$$

where

$$H(Y/X) = -\sum_x \sum_y p(x,y) \log p(y/x) = -\sum_x p(x) \sum_y p(y/x) \log p(y/x),$$

$$H(X/Y) = -\sum_x \sum_y p(x,y) \log p(x/y) = -\sum_y p(y) \sum_x p(x/y) \log p(x/y)$$

are conditional entropies of stimuli and responses.

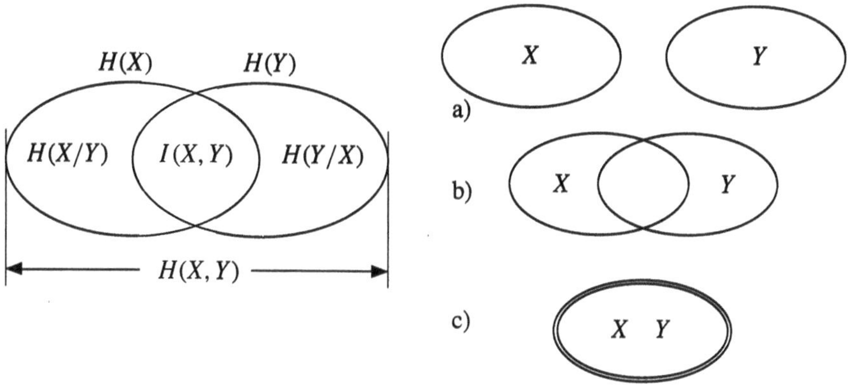

Fig. 1.1

These items reflect the presence of noise in the «informational channel», errors in responses on stimuli. We can name $H(X/Y)$ the «uncompensated entropy of stimuli» and $H(Y/X)$ — the «entropy of errors in responses».

Fig. 1.1 illustrates the interrelations between these notions. The ovals X and Y are picture sets of stimuli and their corresponding responses; the area of the oval represents the unconditional entropy of these variables, the area of intersection is just the mutual information $I(X,Y)$, and the remaining area — the conditional entropy. The area of the whole figure is the joint entropy $H(X,Y) = H(X) + H(Y) - I(X,Y)$.

We must not treat mutual information merely as any messages that the system receives from the environment. It is a very particular case. In general, mutual information is a statistical measure of interaction between two variables. The closer and stronger the interaction, the more mutual information. The other well-known statistical measure of interaction between two variables is the correlation coefficient. Both these measures are closely connected.

Mutual information characterizes the «fitness» of the organism for the environment. Usually for a well-adapted organism this amount of information is so large that the biologist, looking at the traits of the organism, can say a lot about the conditions of its environment. Thus the well-adapted organism is a very «transparent» channel of information; looking at its output we can see what is happening at the input.

When Shannon (1949) introduced his measure of information, he based his work on the concept of probability. It was related to the nature of those problems in the technology of communication that became the departure point of the information theory. But later Kolmogorov (1987) has shown that variable $p(x)$ does not have to be a probability. It can be a «relative part», a «frequency», a «concentration» of the value x among the other values. This interpretation considerably expands the sphere of applications of the information theory. In particular $p(x)$, of course, can be the probability — if we operate with the accidental variable or can introduce (even artificially) any chance mechanism.

For instance: we intuitively feel that the white point against the black background is a very informative object — because the relative area of the point is smaller than the background area. But we have no reason to treat this relative area as a probability until we introduce any mechanism of accidental choice. For example, we throw a little grain of sand and it hits — in an accidental manner — sometimes the white point, sometimes the black background. The probability of this hit will be equal to the relative area. Such mechanism — more or less artificial — can almost always be constructed. Therefore it will not be a mistake if we name $p(x)$ a probability. But we repeat: although such an opportunity exists, the necessity is absent. In other words, Kolmogorov's approach gives us more freedom and permits us to either use the concept of probability, or do without it.

We shall consider x, y discrete variables. But the step of discretization is small enough to replace increments $\Delta x, \Delta y$ by differentials dx, dy, if necessary. We shall use subscripts $(0, 1, 2, i, j, \ldots)$ to distinguish the different gradations of variables.

In general x, y are vectors which have n_x and n_y components correspondingly. (Usually n_x is not equal to n_y). We will use superscripts $(0, 1, 2, k, l, \ldots)$ to distinguish the different components of the vectors.

As soon as variables x and y are discrete we can picture the probability distribution $p(x, y)$ as a matrix with probabilities $p(x_i, y_j)$ in the cells of the matrix. For example, in Table 1.1 we see a matrix of probabilities $p(x_i, y_j)$ where x is a vector of two binary components (x^1, x^2). This vector has four values: $(x_0^1, x_0^2), (x_0^1, x_1^2), (x_1^1, x_0^2), (x_1^1, x_1^2)$. Response y is a scalar variable and has three values:

x / y	(x_0^1, x_0^2)	(x_0^1, x_1^2)	(x_1^1, x_0^2)	(x_1^1, x_1^2)
y_1	0.00	0.01	0.01	0.36
y_2	0.28	0.01	0.13	0.00
y_3	0.17	0.00	0.01	0.02

Table 1.1

We often meet a particular, but simple, important and wide-spread type of variables — *binary variables* — which have only two gradations. For example, white and black, yes and no, true and false, 1 and 0, presence and absence and so on.

At first let us look at a system that contains only one binary element Y. It can be a magnetic dipole («spin») which can be oriented in two directions «up» (y_1) or «down» (y_0); or a «neuron» with two states, «excitation» (y_1) or «rest» (y_0).

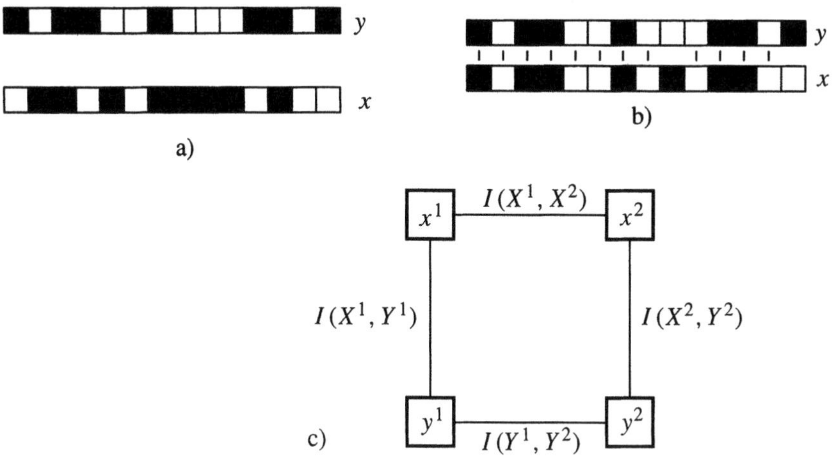

Fig. 1.2

Under the influence of any stochastic force («thermal noise») it can change the state from y_0 to y_1 and back again. In the absence of interaction with any external system, both states are equal in probabilities. The temporal consequence of states of the element is shown in Fig. 1.2a. Here the black cell pictures the state y_0, the white one — the state y_1. The entropy of the ensemble (the set) of these states is maximal and equals $H(Y) = \log_2 2 = 1$ (bit).

Now let any other binary element x (another dipole, another neuron, etc.) exist. This element is the simplest model of the «external environment». The ensemble of the states of these two elements is shown in Fig. 1.2a. If these elements exist independently (any interaction between them is absent) then the entropy of the system is the sum of the entropy of these elements:

$$H(X,Y) = H(X) + H(Y). \tag{1.3}$$

Let there now be an interaction between the elements. For instance, we approximate dipoles to each other (Fig 1.2b), or connect neurons by means of any nervous link. Then the entropy of the system diminishes:

$$H(X,Y) \leq H(X) + H(Y). \tag{1.4}$$

This means that mutual information between x and y arises:

$$I(X,Y) = H(X) + H(Y) - H(X,Y) \geq 0. \tag{1.5}$$

Now, looking at one element, we can form an opinion about the state of the other one.

From this moment a spontaneous behaviour of the system can be directed to decrease the entropy and increase the mutual information. If any random fluctuation or any external influence places the system in a state, where the entropy $H(X,Y)$ is maximal, the system will spontaneously return to the state with less entropy and more mutual information. This process can be treated as: 1) self-organization — spontaneous decrease of entropy; 2) adaptation of the element y to the environment x.

Now we want to consider a more complicated object consisting of two elements, y^1 and y^2, and a more complicated environment consisting also of two elements, x^1 and x^2. Let us suppose that x^1 interacts with y^1 and x^2 — with y^2. Let us suppose also that x^1 interacts with x^2 and y^1 — with y^2 (see Fig. 1.2c). Then we have three types of mutual information:

$I(X^1,Y^1), I(X^2,Y^2)$ — «external» mutual information between the object and the environment;

$I(Y^1,Y^2)$ — «internal» mutual information (correlation, redundance) of the object;

$I(X^1,X^2)$ — «internal» mutual information (correlation, redundance) of the environment.

The more complicated object consisting of n elements contains the same types of mutual information — external and internal.

As a rule, we shall consider the stimulus x to be a «given», «fixed» variable, the responses y to be a «free», «controlled» one.

Generally, the stimulus x is a vector of n_x binary components. It can be pictured in a raster X consisting of n_x cells (Fig. 1.3). Each cell represents one binary component x^k of the vector and can be in one of two states: black — x^k_0 and white x^k_1. In other words, the stimulus is a kind of black-white pattern (picture) in the raster. The response is another binary vector y, which can be represented by another black-white pattern in another raster Y. One of the axes of the raster can be interpreted as time, so then the whole picture presents the consequence of the patterns in time.

If the states of the cells in both rasters are independent, the mutual information between x and y is zero. But if the cells of the raster X interact with the cells of the raster Y then some mutual information between these vectors exists.

A particular, simple, but important case is when picture y reproduces (more or less exactly) picture x; then the mutual information $I(X,Y) > 0$. The process of adaptation in this case looks like a more and more exact reflection, reproduction of x in y, concretization, and «enrichment» of the response y by the new features of x. This means the growth of mutual information between x and y. The maximum of information is achieved when the numbers of degrees of freedom in both rasters are equal and the corresponding cells are in the same states: $x^k = y^k$, i.e., each picture y exactly reproduces the corresponding picture x. In this case the «entropy of errors» $H(Y/X) = 0$ and $I(X,Y) = H(X) = H(Y)$. Any error in reproducing

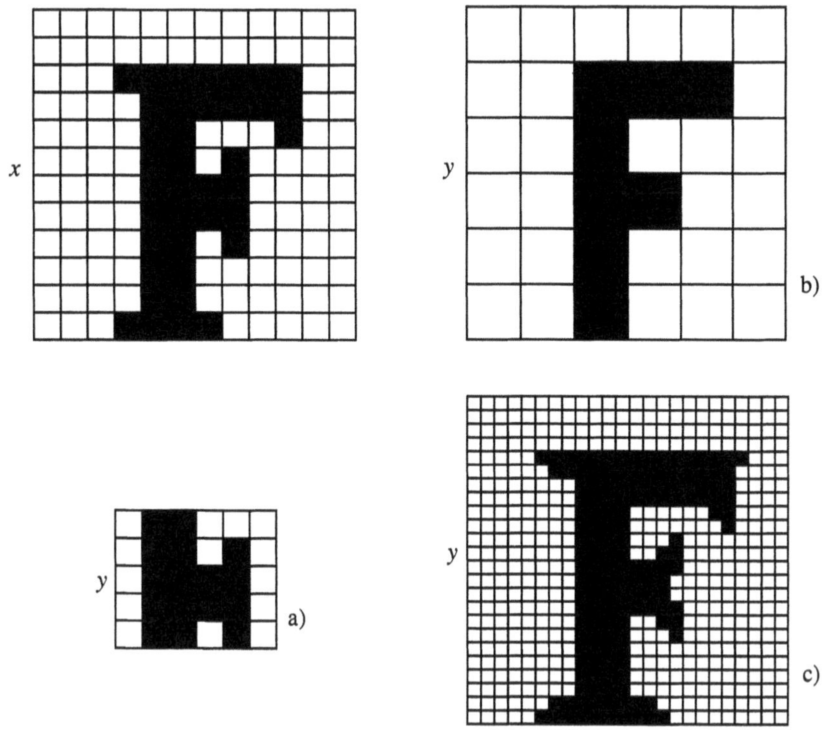

Fig. 1.3

x by y causes an increase of $H(Y/X)$ and decrease of $I(X, Y)$. In particular, if the number of cells in the raster Y is less than in the raster X, errors of reproduction become inevitable and the information between x and y diminishes.

So picture y in Fig. 1.3a is an incomplete reproduction of picture x because the area of the raster Y is too small and y can reproduce only a fragment of x. Picture y in Fig. 1.3b reproduces picture x only very roughly, because the cells of the raster Y are too large. In both cases the information between X and Y is less than the maximal possible one. In Fig. 1.3c the number of cells is large enough and the reproduction of x by y is better than in cases a and b.

On the other hand, if the number of degrees of freedom in the raster Y becomes more than in X, this also causes a decrease of the mutual information. Indeed, if the entropy $H(Y)$ becomes more than $H(X)$ it means that «output» y of the informational channel contains more self-information than «input» x. But this «additional» information (which is absent in x!) can be interpreted only as a «noise» in the channel. And any noise decreases the mutual information.

This obvious binary model will be used as an illustration of general regularities.

1.2 Valuable information

Now we must consider the important problem of valuable (useful) information. This topic has been discussed many times, but the problem remains unsolved: what is a quantitative measure of information value?

It is known that not all information received by the system is useful for it. And for the economy of resources one would transmit, code, store, etc. only useful information. How can one select this information? At first one must have a criterion which permits one to distinguish between useful and useless information.

A large number of such criteria has been suggested in various works. One of the first of them was the «probability of aim achievement» by Kharkevich (1962). Stratonovich (1975) considered energy (and more generally — any resource) a valid criterion.

We must notice that the problem of value in general is a problem of economics and philosophy. Value arises during the process of change (of one thing to another one) and is a relative notion. If one changes a quantity Δx to the quantity Δy, then the ratio $\Delta x/\Delta y$ is the value of one unit of y expressed in terms of x. So we may say that x is the criterion of value for y. On the other hand, the opposite ratio $\Delta y/\Delta x$ is the value of one unit of x expressed in terms of y, so that y is the criterion of value for x.

One of the most general criteria of value in our lives is money. It plays the role of the common change equivalent. But in biology this criterion makes no sense. Usually a biologist uses such criteria as «adaptation», «survival of the species», etc. to estimate the value of any traits or features of a living organism. In principle, this criterion can also be used for human lives: we often regard the survival of mankind as the highest value. But this criterion is not «operational»; we have no procedures to trace all the consequences of each event (particularly information) and its influence on the survival of the species. And, therefore, we cannot estimate the value of this information. Besides, in many cases, the criterion of survival is not adequate. There are many other, particular aims that are not directly correlated to survival. In these cases, the value of information is defined by these aims.

We want to propose an operational criterion for information value which will permit us to compare the values of various types of information.

Let us introduce an «ideal object» similar to ideal objects of physics, such as the «ideal gas», the «perfectly rigid body», etc. It will be a «perfectly adapted system». This means that the system succesfully solves the problems arising in its environment. Really there are no such systems, but a real organism can be very well adapted and in this way approaches this ideal.

Let us denote the responses of an ideal, well-adapted system as $\bar{Y}$. Then

$$I(X, \bar{Y}) = H(X) - H(X/\bar{Y}) \tag{1.6}$$

is the mutual information between the stimulus x and the response y in the ideal system. We can present the variety of stimulus $H(X)$ (i.e., the self-information contained in the variable x) as:

$$H(X) = I(X, \bar{Y}) + H(X/\bar{Y}). \tag{1.7}$$

$I(X, \bar{Y})$ can be interpreted as a *variety* of stimuli which has been compensated by the responses of the system; and $H(X/\bar{Y})$ — is an «uncompensated» variety of stimuli. And we may say that the more the uncompensated variety of any stimulus in an ideal, perfectly adapted system, the less the value of this stimulus for the system.

We often intuitively judge the value of any quantity on the basis of this criterion. We know that if we allow any quantity to vary within a large range, then this quantity is not important, not valuable for us, its variations don't prevent us from achieving our aims. Correspondingly, information about this variable is useless and can be ignored.

Let us consider some examples.

1. In mechanics we come across the notion of the «ignored» or «circular» variable which can be considered a synonym of useless quantity. The ignored variable q is defined as the variable which does not enter Lagrangian L (the function of variables which determines the motion of a system). So variations of this variable do not influence Lagrangian L and, as consequence, do not influence the state of the system. The range of variations of this variable may be unlimited. It is clear that a variable with an unlimited range of variations is not a useful variable and information about this variable is not valuable. Such a variable can be excluded from the equations describing the system. But the notion of the ignored variable permits us to distinguish only two gradations of value — to be, or not to be, ignored. And the range of variations can only be of two types: limited or unlimited. In other words, it is a qualitative estimation. Through using the entropy notion we can introduce a more adequate quantitative measure for the range of variations and obtain a more exact measure of value.

2. In sociology and economics there is a method to estimate the «importance» and «insistency» of various needs. The higher the variance in the consumption of any object, the less important this object becomes to the consumer. So variance in the consumption of bread is lower than in the visitation of the cinema. Hence bread is a more important (we should say — more valuable) object of consumption than the cinema. And, consequently, information about bread is (on average) more valuable than information about the cinema.

3. Standard deviation σ of the concentration in blood of man for pCO_2 is equal to 5.4%, for pH — 0.05%. In the case of normal distribution, entropy H is connected with standard deviation σ through the expression:

$$H = \log \sqrt{2\pi e}\, \sigma = \log 4.13\sigma \tag{1.8}$$

Hence the uncompensated variety $H(X/\bar{Y})$ for the variable pCO_2 is more than for the variable pH at $\log(5.4/0.05) = 6.5$ bits. We may say that under other equal conditions variable pH contains more useful information than the variable pCO_2 at 6.5 bits. Every doctor knows that pH contains more useful information about the state of the patient than pCO_2.

4. We can *normalize* variable x, and introduce a «natural scale» for this variable so that the *origin* will coincide with the average value of x, and the *unit* will be equal to the standard deviation σ in a well-adapted system. Expressed in terms of normalized variables a conditional, «uncompensated» entropy $H(X/\bar{Y})$ becomes zero in an «ideal», well-adapted system. Then the valuable information will coincide with the mutual information for the normalized variables.

5. If it is our purpose to recognize a letter, we can be satisfied with picture y in Fig. 1.3b which contains some errors in the reproduction of Figure x. But many details in figure x are superfluous and can be omitted in picture y. These details are the useless information $H(X/\bar{Y})$ in picture x.

Thus we can say: not all information contained in x is useful and must be represented in y. Any stimulus contains many details which are useless for the system during the given task. On the other hand, not all errors in responses are fatal and lead to system failure. In other words, in a well-adapted system the uncompensated entropy $H(X/\bar{Y})$ (like the entropy of errors $H(\bar{Y}/X)$) must not always be zero.

Hence $H(X/\bar{Y})$ in a well-adapted system is a measure of «useless» information, $I(X,\bar{Y})$ is the useful information, and $H(X)$ is the full information contained in x. In other words, full (self) information contained in x is the sum of the useful and the useless information. The quantity $H(X/\bar{Y})$ can be considered as the *origin* for the valuable information.

In short, the useful information, contained in x coincides with the mutual information $I(X,\bar{Y})$ in a well-adapted system. Hence we don't need a special quantitative measure of information value; its measure is the mutual information itself, if variable x is normalized and entropy $H(X/\bar{Y})$ is the origin for information.

1.3 Principle of the information maximum

Now we can formulate the main postulate of this work — the principle which permits us to explain many facts and regularities of behaviour and the evolution of various types of systems — from physical bodies to living organisms and social organizations. It is the *principle of the information maximum* (Golitsyn, 1984; Golitsyn and Petrov, 1990, 1991):

$$I(X,Y) = \sum_x \sum_y p(x,y) \log \frac{p(x,y)}{p(x)p(y)} = \max . \tag{1.9}$$

This principle means: *the system aspires to choose such response y which provides the maximum valuable information about the given stimulus x.*

As a rule, there are some restrictions which prevent the system from reaching the absolute (unconditional) maximum of information. The system has to be satisfied with the conditional maximum. A very general and typical condition is the restriction of the average resource $R(X, Y)$:

$$\sum_x \sum_y p(x,y) r(x,y) = R(X,Y) \tag{1.10}$$

where $r(x, y)$ is a resource expense in state (x, y).

We consider resources to be energy, materials, time, space, number of operations, etc. Usually there are several restrictions so that we must write several expressions like (1.10).

We must also take into account the condition connecting probabilities:

$$\sum_x \sum_y p(x,y) = 1. \tag{1.11}$$

Equations (1.9), (1.10), and (1.11) can be joined by means of the Lagrange multipliers α and β and presented in the form:

$$\begin{aligned} L &= \sum_x \sum_y p(x,y) \log \frac{p(x,y)}{p(x)p(y)} - \alpha \sum_x \sum_y p(x,y) \\ &\quad - \sum_k \beta^k \sum_x \sum_y p(x,y) r^k(x,y) = I(X,Y) - \alpha - \sum_k \beta^k R^k(X,Y) \\ &= H(X) - H(X/Y) - \alpha - \sum_k \beta^k R^k(X,Y) \\ &= H(Y) - H(Y/X) - \alpha - \sum_k \beta^k R^k(X,Y). \end{aligned} \tag{1.12}$$

A physical sense of the multiplier β is a deficit of resources. So in thermodynamics $\beta = 1/T$ where T is an absolute temperature — the average energy of one degree of freedom. The less energy the system has, the higher the β. In other cases β may be a deficit of space, time, power, etc. The higher the deficit, the higher the weight β and the significance of the corresponding resource in the total sum. If the deficit of the resource R^k is absent, then β^k is equal to zero and the corresponding item in the total sum can be omitted.

As a rule we shall consider only one resource in this work and shall interpret it as energy. It is the most important amongst the resources and many others (number of operations, time, etc.) can be reduced to it.

Let us look at the example of the adaptation and the increasing of information — the formation of the conditional reflex in the rat. The experiment looks like this: when the lamp flashes at the right side of the experimental box, the rat has to press the right key; if the lamp flashes at the left side, the rat has to press the left key. When correct, the rat gets an reinforcement. In this example variable x (stimulus) is the light. It has two gradations: x_0 — the left side, x_1 — the right side. For reasons of simplicity both states have equal probabilities: $p(x_0) = p(x_1) = 0.5$. Then the entropy of this variable is

$$H(X) = \log 2 = 1 \text{ bit}. \tag{1.13}$$

Variable y in this example is the response of the rat: y_0 — pressing the left key, y_1 — the right key. Let these responses have equal probabilities also. Then the entropy:

$$H(Y) = \log 2 = 1 \text{ bit}. \tag{1.14}$$

Until the rat has learned the connection between the stimulus and the response, the mutual information between them is zero $I(X,Y) = 0$ and the entropy of the system «stimulus + response» equals:

$$H(X,Y) = H(X) + H(Y) = 1 + 1 = 2 \text{ bits}. \tag{1.15}$$

This situation is pictured in Fig. 1.1a. Set X and set Y do not share an intersection and the mutual information between these sets (the measure of intersection) is zero.

In the process of learning (formation of the conditional reflex) the mutual information $I(X,Y)$ increases and the entropy $H(X,Y)$ correspondingly decreases. This is presented in Fig. 1.1b as the intersection of sets X and Y. At the end of the process of learning, the stimulus and the response will be connected rigidly: where the light is, the rat appears. Both variables behave as one (full coincidence of set X and set Y, see Fig. 1.1c). And it means a decrease in the number of degrees of freedom and the entropy of the system $H(X,Y)$ to 1 bit. Accordingly, the information in the system increases. The structure of the system is complicated, the system acquires one new connection — the conditional reflex.

The amount of the mutual information

$$I(X,Y) = H(Y) - H(Y/X) \tag{1.16}$$

characterizes the degree of adaptation of the rat to the situation, where $H(Y)$ is variety and $H(Y/X)$ — unexactness of the responses.

Another example: I translate a Russian text into English. In this case the stimuli x are Russian words, the responses y — their English equivalents, $H(X)$ — the variety of the Russian vocabulary, $H(Y)$ — the variety of my English vocabulary, $H(Y/X)$ — inexactness of the translation, — etc. A skill of the translator (degree of «adaptation» to his profession) is characterized by the amount of mutual information $I(X,Y) = H(Y) - H(Y/X)$ where $H(Y)$ characterizes variety and $H(Y/X)$ — inexactness of the translated text.

These examples must demonstrate how the principle can be interpreted when applied to various concrete situations.

The information maximum principle is a postulate. And a postulate can not be proven. Its correctness is verified by the coincidence of the consequences with the experience. We shall give some further examples of such a coincidence. But now we want to make some remarks which cannot be considered as proofs, but as mere heuristic foundations of this principle.

The principle of the information maximum has deep evolutionary roots. Any physical interaction leads to the increase in the mutual information. The evolution of the physical systems was accompanied by the complication of their structures and by the growth of the information contained in these structures. Some authors (Barrow and Tipler, 1982; Davies, 1982; Breuer, 1984, etc.) note a wonderful coincidence of many physical and cosmological constants which tell us about a very precise «tuning» of the physical Universe, i.e., about the great amount of information concluded in the structures of the Universe.

This process was continued in biological evolution, where the adaptation, the natural selection, and the rise of new species means the accumulation of the new information. One of the most reliable criteria of progress in the biological evolution is the increase in the mental level of living creatures. Morphologically it is expressed through «cephalization», the development of the nervous system, its growth, complication, and centralization. But what is the nervous system? It is a device intended for the perception, storage, and processing of information.

The tendency to increase and accumulate information continues to play a role in human history and finds an expression in the development of language, art, science, technique, and other cultural phenomena. Now, that we've entered the «information society», this tendency becomes even more obvious.

Thus with the concept of the information maximum we obtain a quantitative measure of progress in Nature and Society, a criterion which permits us to compare the higher and the lower forms of life and to predict (in a very general outline, of course) the further direction of evolution.

As is clear from the principle of the information maximum

$$L(X, Y) = H(Y) - H(Y/X) - \alpha - \beta R(X, Y) = \max \qquad (1.17)$$

there are three main tendencies in the behaviour and evolution of systems:

1. *Expansion* — the aspiration to increase the number and the variety of the states of the environment $H(X)$ in which the system can exist. This is made possible by the corresponding increase in the number and the variety of the system responses $H(Y)$. We name this tendency «search behaviour».

2. *Idealization* — the aspiration to increase the «exactness» of the responses, i.e., to decrease the uncompensated entropy of the stimuli $H(X/Y)$, as well as the entropy of the errors $H(Y/X)$.

3. *Economy* of resources. This is expressed on the one hand in the choice of combinations (x, y) with minimal resource expense $r(x, y)$; on the other hand

— in the aspiration to decrease the deficit of resource β, i.e., to increase resource supply.

Here we want to look at the example of the evolution of one cultural subsystem.

One of the most powerful and universal subsystems of culture is language. We shall consider it in detail in Chapter 3. But here we will look at a particular, but important kind of symbolic systems — the scale of numeration. In this case x («stimulus») is interpreted as any quantity and y (response) — as its representation in the system of numbers. The principle of the information maximum (1.17) requires the system to be able to represent the quantities

a) with the maximal possible completeness ($H(Y) = \max$);

b) with the maximal exactness ($H(Y/X) = \min$);

c) with the minimal expence of resources (energy, time, place, number of signs and operations, etc., $\beta R(X, Y) = \min$).

The evolution of the scale of numeration shows us all three tendencies: the growth of the range and variety of responces $H(Y)$, the decrease of the diapason of errors $H(Y/X)$, and the economy of resources $R(X, Y)$. Let us give some examples.

Many tribes in Australia until the 19th century used only two numbers: one and two. Combinations of these (two-two-one, etc.) made it possible to construct numbers up to six. All the larger quantities were simply designated as «many». On the one hand we see a very restricted «variety of responses» and on the other hand — a very low exactness in the representation of quantities that are more than six or less than one. Similar situations were typical for the primitive stage of development of many other peoples.

Further development everywhere was directed toward the enlargement of the set of numbers to finally establish the natural scale of numbers including null and infinity. But at this point evolution did not finish and the diapason of numbers continued to grow, including negative, imaginary, complex, and hypercomplex numbers. On the other hand the exactness increased also — thanks to the use of fractions — simple, complex, rational, irrational, continued, and so on.

The other stable tendency of this evolution was an economy of the resources. Everywhere, the first (and the less economic) methods to fix the quantities were to bend fingers, to tie knots, to put aside pebbles or sticks, to make incisions, and so on. In all these methods the number of signs was equal to the number of objects signified. But these methods were not only non-economic, but not obvious. It is known that man can take in only about 7–8 objects in one glance (see also Chapter 3). This is the reason why we unite objects in groups and count the groups, the groups of groups, and so on. There arises a hierarchy of quantities and this hierarchy receives a representation in the system of numeration. In this system we obtain the opportunity to define in one single symbol not a single object, but a group of objects. And groups of different levels of the hierarchy are defined by symbols, which are distinguished either through form (Roman numerals), or through position (Arabic numerals). These methods permit us to sharply contract the number of signs and, correspondingly, the other resource expences.

One more general argument confirming the principle of the information maximum is that this principle is the further development of some famous physical and biological principles such as the principle of the entropy maximum, the method of the maximum likelihood, the Curie principle, the principle of homeostasis, etc.

For instance, if resource R is energy and items $H(Y)$ and $H(Y/X)$ in (1.17) are fixed, we obtain the principle of the economy of resources (3). Thus (3) is a particular of the information maximum principle. This economy becomes of prime importance also, when the resource deficiency β is too large and the last item in (1.17) predominates over the two others.

We must also point out the famous «law of requisite variety» by Ashby (1956). This law suggests that the variety of responses of the organism (which can be characterized by the entropy $H(Y)$) must be equal to, or higher than, the variety of the stimuli (conditions of the environment)$H(X)$. It is the necessary condition for survival of the organism in this environment. Necessary, but not sufficient. The other condition is the «exactness» of the responses, the correspondence of these responses to the stimuli. The information maximum principle is a combination of these two conditions.

These principles can be perceived as particular cases of the information maximum principle. Let us consider some of these principles in detail.

1.4 Expansion: Principle of the entropy maximum

If item $H(Y/X)$ in expression (1.17) is fixed we obtain the *principle of the entropy (self-information) maximum.* This principle describes a tendency to increase the number and the variety of the system responses $H(Y)$ («expansion»).

At first the principle of the entropy maximum was formulated in physics. But now it is broadly applied to many areas of life (including human activity) such as biology, economics, linguistics, sociology, etc. The modern usage of this principle is based on works by Jaynes (1957), who considered the informational entropy a subjective measure of our unknowledge. His principle of the entropy maximum is a generalization of the famous principle of «equirandomization of probable events»: if we know nothing about variable y, we must suppose that all values of this variable have the same probabilities: $p(y) = \text{const}$. But this result can be deduced from the principle of the entropy maximum in the particular case of absence of any supplement restrictions. It was normal to generalize this principle in the case where such restrictions were present.

If we know anything about the restrictions of variable y (such as the average of y, the variance of y, etc.) we can obtain a more realistic probability distribution, through adding these restrictions, as supplement conditions, to the principle of the entropy maximum. A very general and typical form of restriction is:

$$\sum_y p(y)r(y) = R(Y). \tag{1.18}$$

The principle of the entropy maximum can be described as follows: if y is an accidental variable and there exists some statistical restriction on any function of this variable, then the probabilities $p(y)$ can be obtained from the condition

$$H(Y) = -\sum_y p(y)\log p(y) = \max_{p(y)} \tag{1.19}$$

taking into account the condition (1.18).

We must also take into account that the sum of the probabilities must be equal to 1:

$$\sum_y p(y) = 1. \tag{1.20}$$

Equations (1.18), (1.19), and (1.20) can be joined by means of the Lagrange multipliers α and β:

$$\begin{aligned} L = H(Y) - \alpha - \beta R = \\ -\sum_y p(y)\log p(y) - \alpha\sum_y p(y) - \\ \beta\sum_y p(y)r(y) = \max. \end{aligned} \tag{1.21}$$

Let us look at some consequences of this principle.

a). How does entropy $H(Y)$ change when the probability $p(y)$ grows from 0 to 1? We shall consider a binary case, so that $p(y_0) = 1 - p(y_1)$. In this case

$$\begin{aligned} H(Y) &= -p(y_0)\log p(y_0) - p(y_1)\log p(y_1) \\ &= -[1 - p(y_1)]\log[1 - p(y_1)] - p(y_1)\log p(y_1). \end{aligned} \tag{1.22}$$

One can distinguish two different stages in this process (Fig. 1.4a):

I. When $p(y_1)$ is zero or near zero, then growth of $p(y_1)$ causes a growth of the entropy.

II. When $p(y_1)$ approximates 1, then the entropy goes to zero, i.e., diminishes. Hence there exists a point (y_1) where entropy is maximal.

b). One can see that there exists an optimal distribution of probabilities $\bar{p}(y)$ which provides the maximum of Lagrangian $L(Y)$, i.e., the maximum of entropy

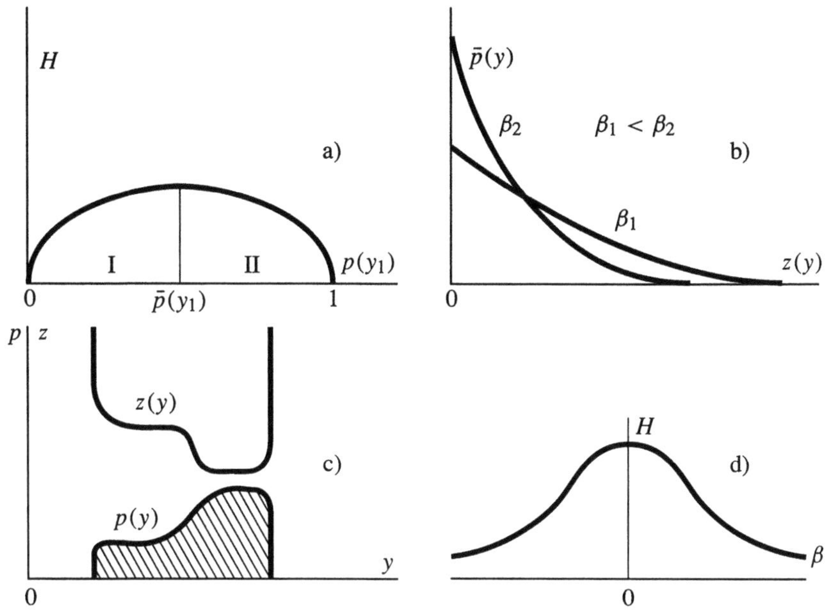

Fig. 1.4

under the given restrictions (1.18) and (1.20). Let us find the partial derivative of L in respect to any probability $p(y)$ and bring it down to zero:

$$\partial L/\partial p(y) = \partial[(-\sum_y p(y)\log p(y) - \alpha\sum_y p(y) - \beta\sum_y p(y)r(y))]/\partial p(y) = -\log p(y) - 1 - \alpha - \beta r(y) = 0. \tag{1.23}$$

Or

$$\log p(y) = -1 - \alpha - \beta r(y). \tag{1.24}$$

Or

$$\bar{p}(y) = \exp[-1 - \alpha - \beta r(y)] = \exp(-1 - \alpha)\exp[-\beta r(y)] = \exp[-\beta r(y)]/Z \tag{1.25}$$

where Z is the so-called «statistical sum» which plays the role of the normalization factor:

$$Z = \exp(1 + \alpha) = \sum_y \exp[-\beta r(y)]. \tag{1.26}$$

We can see from (1.25): the more $r(y)$, the less $p(y)$. For example, let $r(y)$ be the resource expense which is necessary to achieve state y. Then (1.25) means: the higher the resource expense the lower the probability of the correspondent state y

(Fig. 1.4b). If the resource expense is infinite, $r(y) = \infty$, then $p(y)$ is zero, i.e., the state y is unattainable. For instance, if we have a potential well with infinitely high walls, then the probability distribution is concentrated inside the well and probabilities will be zero outside the well (Fig. 1.4c).

c). Now let us pay attention to the role of the multiplier β which characterizes a deficit of resources. The higher the β, the closer the distribution $p(y)$ is concentrated around zero (Fig. 1.4b), the higher the uneveness of distribution, and the lower the entropy (variety) of responses $H(Y)$. (The entropy is maximal when the probability distribution is even). In other words, a deficit of resources leads to a decreasing variety of responses of the system.

We can find an expression for the dependence of H on β for the particular binary case. In this case there are two levels of resource expense. Let their values be $r(y_0) = 0$ and $r(y_1) = r$. Let us substitute the probability $p(y)$ in the expression for entropy (1.22) by the expression (1.25):

$$\begin{aligned} H &= -\sum_y p(y) \log p(y) = \frac{1}{Z} \sum_y \exp[-\beta r(y)] \beta r(y) \\ &- \log Z = \frac{1}{Z} \exp(-\beta r) \beta r - \log Z. \end{aligned} \tag{1.27}$$

In a binary case Z is equal to

$$\begin{aligned} Z &= \sum_y \exp[-\beta r(y)] = \exp[-\beta r(y_0)] \\ &+ \exp[-\beta r(y_1)] = 1 + \exp(-\beta r). \end{aligned} \tag{1.28}$$

Hence

$$H = \frac{\beta r}{1 + \exp \beta r} + \log[1 + \exp(-\beta r)]. \tag{1.29}$$

The dependence of H on β is pictured in Fig. 1.4d. One can see that the entropy is maximal $H = 1$ when $\beta = 0$. A deficit of resources ($\beta > 0$) leads to a decrease in entropy.

But if the number of gradations m of the variable y is limited ($m < \infty$, particularly $m = 2$) then the deficit β can be negative («surplus» of resources, $\beta < 0$). In this case the left branch of the curve $H(\beta)$, symmetrical to the right one, also exists (Fig. 1.4d). Thus a decrease in resources, as well as an increase in them, causes a decrease of the entropy. In physics this effect is known as *negative absolute temperature*. In psychology and ergonomics this regularity is known as *the «Yerkes — Dodson law»*: the maximal effectiveness of any activity (which is conditioned by the maximal variety of responses $H(Y)$) is achieved on the *average* level of activation (Yerkes and Dodson, 1908).

d). One of the consequences of the principle of the entropy maximum is the famous Zipf's law (Zipf, 1949).

Very often the more experience you have in executing any activity, the easier it is to create something new. Formally this can be expressed as the inverse dependence between the increment of the resource expense dr and the total experience y:

$$dr = ady/y. \tag{1.30}$$

For instance, the more papers y that have been written by a scientist, the easier it becomes for him to write one more paper, so that the resource expense on a new paper is inversely proportional to the number of papers written.

Integrating expression (1.30) we have:

$$r(y) = a\log y + c. \tag{1.31}$$

Substituting (1.31) in (1.25) we obtain a probability distribution

$$p(y) = \exp[-\beta(a\log y + c)]/Z = by^{-\beta a} \tag{1.32}$$

where $b = \exp(-\beta c)/Z$.

It is merely Zipf's law: the more y, the less the probability $p(y)$ (or the relative part, relative frequency) of state y. In logarithmic axes this dependence turns out to be linear with the negative inclination:

$$\log p(y) = -\beta(a\log y + c) - \log Z. \tag{1.33}$$

Hence Zipf's law is the optimal distribution (from the point of view of the entropy maximum) under the condition (1.30).

e). Let us consider a particular case when $r(y)$ is constant and does not depend on $y : r(y) = \text{const}$. As follows from the expression (1.25), the probability $p(y)$ will also be constant and independent of y. In this case the optimal probability distribution, providing entropy maximum, is the uniform and equal $p(y) = 1/m$, where m is the number of gradations.

f). There exists an important problem about the optimal structure of the system. This problem arises in the theory of communication, psychology, linguistics (see Chapter 3), etc.

Let the response be described by n independent components of a vector («degrees of freedom») and each component have m gradations of equal probabilities. Let the resource expenses be proportional to the number of gradations

$$r = am. \tag{1.34}$$

The total resource expense is restricted by value

$$R = amn. \tag{1.35}$$

Hence

$$n = R/am. \tag{1.36}$$

What is the optimal number of gradations providing maximal «effectiveness» of the response, i.e., the maximal total information I in respect to the given resource expense R. We can write for I

$$I = I(m)n = I(m)R/am \tag{1.37}$$

where $I(m)$ is the information in one degree of freedom.

If $p(y) = \text{const}$, then $I(m)$ is equal (in natural units) to:

$$I(m) = H(Y) = -\sum_y p(y) \ln p(y) = \ln m. \tag{1.38}$$

Substituting $I(m)$ in expression (1.37) we obtain:

$$I = \ln mR/am. \tag{1.39}$$

To find an optimal value m, the derivative of I in respect to m must be brought down to zero:

$$dI/dm = R(1 - \ln m)/am^2 = 0. \tag{1.40}$$

Hence

$$\ln m = 1 \tag{1.41}$$

and

$$m = e = 2.71. \tag{1.42}$$

In so far as y is a discrete variable, m can be the integer number closest to 2.71, i.e., can be equal to 2 or 3.

There are many examples of manifestations of this conclusion. Linguists have noticed that natural classifications dividing objects into two or three classes are very wide-spread in language and make up a majority of all classifications. So we divide things into big and small, black and white, hot and cold; or into big, medium and small, black, gray and white, hot, warm and cold — and so on.

The other example is the «all or nothing» law which characterizes many biological and physiological processes: often a living system either reacts to the stimulus with maximal intensity, or does not react at all, i.e., response has only two gradations.

Binary code in computers is the third example. This code was chosen because it proved to be the best and easiest for technical performance.

One more interesting case, where dependence $I(m)$ is distinguished from (1.39) and thus allows us to obtain such aesthetic constants as the «golden section», consonances of the musical scale, etc. will be discussed in Chapter 5.

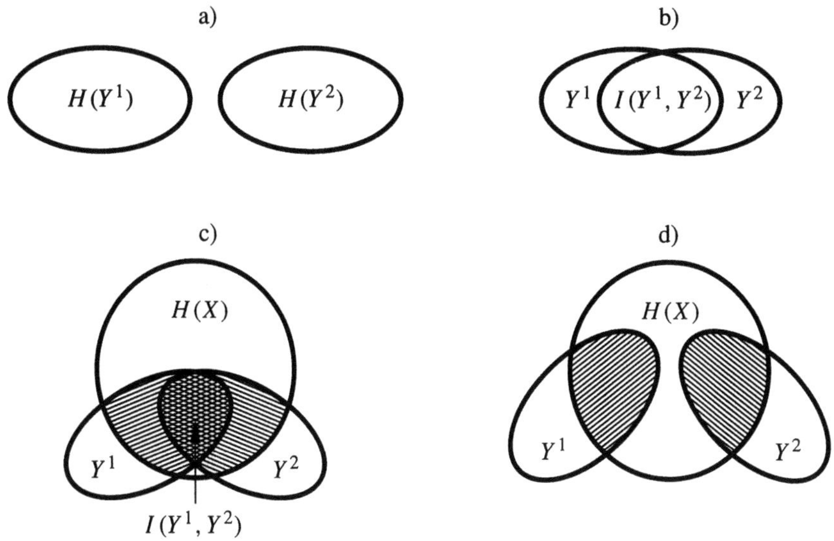

Fig. 1.5

So, many interesting empirical regularities and constants can be deduced from the principle of the entropy maximum. We shall use some of the results obtained in our further consideration.

g). Let us consider two variables y^1 and y^2 which characterize the responses of the system. If these variables are statistically independent, i.e., the correlation between them is absent, then ovals representing these variables (see Fig. 1.5a) do not intersect each other. Total entropy (variety of responses) $H(Y) = H(Y^1, Y^2)$ is equal to the sum of two entropies — $H(Y^1)$ and $H(Y^2)$. It is represented by the two areas of ovals.

Let us assume a correlation between y^1 and y^2 exists. It can be pictured as the intersection of two ovals (Fig. 1.5b). The area of this intersection represents the internal mutual information $I(Y^1, Y^2)$ between y^1 and y^2 which is the measure of correlation between these variables. Total entropy of system responses is diminished by this information:

$$H(Y) = H(Y^1) + H(Y^2) - I(Y^1, Y^2). \tag{1.43}$$

We can see that the total variety of the system responses is greater if the correlation («internal mutual information») between the responses is less. This effect can be used by the system for its more perfect adaptation to the environment.

If the indeterminance of the environment (variety of stimuli $H(X)$) is too large (more than the variety of responses $H(Y)$) then, as a rule, decorrelation is useful; it can lead to the growth of mutual information between the system and environment. Substituting (1.43) in the expression for external information (1.12) we obtain:

$$I(X,Y) = H(Y) - H(Y/X) = H(Y^1) + H(Y^2) - I(Y^1,Y^2) - H(Y/X). \quad (1.44)$$

If $H(Y^1) + H(Y^2) - H(Y/X)$ is fixed, then one can see that the less internal information there is between responses $I(Y^1, Y^2)$ («redundance»), the more external information $I(X,Y)$ exists between the system and the environment. An illustration of this situation is pictured in Fig. 1.5c,d. Particularly, if the stimuli x^1, x^2 are not correlated, then decorrelation of corresponding responses y^1, y^2 is useful as soon as it provides a more perfect adaptation to these stimuli (see Fig. 1.2 c).

Example: in Fig. 1.3 the large size of cells in the raster Y is the simplest form of correlation between the variables y^k: some neighbouring cells are always in the same state and can be considered as one cell of the larger size («domain», «block»). Decorrelation is presented as the division of the large cells («blocks») in raster Y into the smaller ones («bricks»). It allows us to reproduce the picture x in picture y with a greater exactness and translate more information from x to y.

h). The more the deficit of the resources β, the more the internal mutual information (correlation, organization) between the elements of the system y^k (magnetic dipoles, neurons, human beings, etc). This means, particularly, that elements are united in groups: magnetic dipoles — into «domains», neurons of the retina — into «receptive fields», human beings — into «organizations», etc. Usually these correlations disappear and groups desintegrate when the resources of the system increase.

As a rule, in the process of evolution the resources of the system increase. Hence, strong correlations between the elements of the system are usually observed in the early stage of evolution. These correlations then diminish, and the number of degrees of freedom and variety of responses increase. This is a «normal», «progressive» course of evolution. But sometimes resources can decrease and one can observe an inverse course of development — «regress».

Let us now consider the evolution of any technical system, for example — that of the wheeled vehicle. Archaeological investigations show that the primitive wheels were fastened to one another with an axle and, therefore, were rigidly connected. Then, the front axle couldn't rotate horizontally, i.e. it was rigidly connected with the corps and by means of the corps — with the second axle. All these connections hardly constrained the number of degrees of freedom and the variety of motions (responses!) of the system. There was, in essence, only one degree of freedom — rotation around the axle. It prevented the system from reacting adequately to the conditions of the environment — bends and holes in the road.

The evolution of the system was directed towards «unbinding», towards the decorrelation of the elements of the system, towards an increase of the number of degrees of freedom and variety of motions of the system. The crucial moments of this evolution were:

- independent rotation of both wheels around the axle;
- rotation of the front axle horizontally in respect to the corps;
- a «trapezium» which permits the wheel to turn horizontally independent of the axle;
- springs — to de-unite the axle and corps;
- independent hanging of the wheels;
- pneumatics — to create a space between the road and the wheels;
- differential — to provide independent rotation of the leading wheels.

All these devices provided a large flexibility and variety of the system's movements y, a large degree of adaptation of the vehicle to the conditions of the road x.

Thus, decorrelation and growth of the number of degrees of freedom n_y and the entropy of responses $H(Y)$ of the system is one of the most important tendencies of development. It provides an increase of the mutual information between the system and the environment and a more perfect adaptation of the system.

Hence «decorrelation» can be useful in itself. But usually decorrelation is only a means to reorganize the system of responses, to establish *new* correlations between the components of the response vector and, as a result, to *obtain a higher number of more complicated structures of responses.* In other words, we divide large blocks into small bricks, but the bricks themselves are not the end, but the means to build more various and complicated blocks. This reorganization is useful because the external environment is not fully randomized, it is more or less organized. And the organization of the responses must correspond to the organization of the environment. Both correlation and decorrelation can be useful, but under different conditions or in different stages of adaptation.

Indeed, if the entropy $H(Y)$ becomes more than $H(X)$, this also causes a decrease of the mutual information. It means that «output» y of the informational channel contains more self-information than «input» x. But this «additional» information (which is absent in x!) can be interpreted only as a «noise» in the channel. And any noise decreases the mutual information. In this case the introduction of additional correlations between the degrees of freedom of the raster Y can «smooth» the noise and improve the reproduction of x by y. Such smoothing, suppressing the noise, is a device frequently used in the technique of communication, image processing, photography (retouch), and so on. One can say that correlations between responses (elements of Y) become profitable in the presence of a noise.

Now let us look at another particular case of the information maximum principle, which describes processes of correlation, organization, and decrease of the entropy.

1.5 Adaptation as organization

Let us compare two expressions — the principle of the information maximum (1.17) and the principle of the entropy maximum (1.21). The most interesting distinction between them is the item $H(Y/X)$ which expresses the tendency to decrease conditional entropy. If we fix all items in expression (1.17) except $H(Y/X)$, then the expression (1.17) becomes the «*principle of the conditional entropy minimum*». This tendency to decrease the entropy can be considered as *organization.* Let us consider this tendency in more detail.

The evolutionary approach arose almost simultaneously in physics and biology. But it was based on different principles and supposed different directions of the evolutionary process. According to the physical principle of the entropy maximum, any reserved system (even the Universe as a whole) aspires to reach an equilibrium, maximal disorganization, the destruction of any structure. In biology, on the contrary, we observe a growth of organization, a departure from equilibrium, a complication of structures. The evolutionary approach in biology must explain how the new, more and more complicated, structures rise. We were confronted with the problem: how can order arise in a world where entropy as a total grows? Why do physical and biological systems develop in opposite directions?

If we didn't take into account vitalistic conceptions, which assume the existence of any sort of «vital force» to explain the peculiarities of living systems, we would have to admit that we have no answer to these questions.

Before going further we must ask some more questions: what do we imply when we speak about a «high regularity», «high organization» of a living system? What facts show us this high level of organization?

About these questions, Blumenfeld (1974) wrote:

«All conversations about the «antientropical tendencies» of biological evolution, about the »exclusive regularity» of living matter are based on an evident misunderstanding. According to the physical criteria, any biological system is no more regulated than a piece of rock of the same weight».

In our opinion, this note by Blumenfeld allows us to draw the conclusion that a «high organization» of a living system, intuitively evident for us, is placed at quite another level and cannot be understood by means of the traditional thermodynamical methods.

Our answer to these questions consists of the following: a high organization of a living system is a high degree of adaptation of this system (as well as its parts) to the conditions of the environment. This adaptation is displayed first of all on the macroscopical level of the organization of the system, which is the most accesible one for our observations. This adaptation can be adequately described as the growth of the mutual information between the system and the environment. This growth includes, particularly, a decrease of the conditional $H(Y/X)$ (as well as joint $H(X,Y)$) entropy. This last effect can be interpreted as organization.

Let us consider this effect in more detail. We must distinguish two types of the behaviour:

1. *Free* or *spontaneous* behaviour when the system aspires to return to the equilibrium (to the optimal state) $\bar{p}(y)$ and to increase its objective function: $p(y) \to \bar{p}(y)$, $L(p) < L(\bar{p})$.

2. *Forced* or *compulsory* behaviour when the optimal, equilibrium state of the system is shifted under the influence of any external factor $x : \bar{p}(y) \to \bar{p}(y/x)$. Objective function $L(\bar{p})$ can decrease in this process.

Let us give a formal description of both types of behaviour.

As soon as the objective function L must *increase* in free (spontaneous) behaviour we can write:

$$dL/dt = \partial L/\partial p \cdot dp/dt \geq 0. \tag{1.45}$$

Hence

$$\operatorname{sgn}\left(\frac{dp}{dt}\right) = \operatorname{sgn}\left(\partial Lp\right). \tag{1.46}$$

Using a linear approximation of this expression we can write:

$$\tau\frac{dp}{dt} = \partial Lp \tag{1.47}$$

where τ is any time constant.

Compulsory behaviour can be described through the restriction (similar to the restriction (1.18)) which is imposed on the system:

$$\sum_{xy} p(y,x)r(y,x) = R(Y,X). \tag{1.18a}$$

Here x is any external factor connected with the state of the system y by any type of interaction. This connection changes the objective function of the system and must be supplied to this function by means of Lagrange multiplier β. Let the initial objective function L be entropy $H(Y)$; the initial equilibrium state — an even probability distribution $(y) = \text{const}$. The new objective function is:

$$L(X,Y) = H(Y) - \alpha - \beta R(X,Y). \tag{1.48}$$

The interaction causes a *compulsory deviation* of the new equilibrium state $\bar{p}(y/x)$ (state conditioned by x) from the old one $\bar{p}(y)$. We can use expression (1.25) to find a new state:

$$\bar{p}(y/x) = \exp[-\beta r(x,y)]/Z \neq \bar{p}(y) \tag{1.49}$$

where $Z = \sum_y \exp[-\beta r(x,y)]$.

If x is changed then the new equilibrium state is changed also. Let x be a random variable with the probability distribution $p(x)$. Then we can find conditional entropy $H(Y/X)$ in the new equilibrium state:

$$\begin{aligned} H(Y/X) &= -\sum_x p(x) \sum_y p(y/x) \log p(y/x) \\ &= \beta R(X, Y) + \log \sum_y \exp[-\beta r(x, y)] \end{aligned} \tag{1.50}$$

and mutual information between x and y:

$$\begin{aligned} I(X, Y) &= H(Y) - H(Y/X) \\ &= H(Y) - \beta R(X, Y) - \log \sum_y \exp[-\beta r(x, y)]. \end{aligned} \tag{1.51}$$

Usually r can be treated as a resource and β — as the deficit of this resource. One can see that the more deficit β, the less the «resistance» of the system to the external compulsion; the more deviation of the new optimal, equilibrium state from the old one, the less conditional entropy (see also Fig. 2.1) and the more mutual information between x and y. If $\beta = 0$ then the new normal state (probability distribution) is even $p(y/x) = \text{const}$ (i.e., coincides with the old one), $H(Y/X) = H(Y)$ and $I(X, Y) = 0$.

Example: Let y be a state of a magnetic dipole in our binary model, and x is the state of any other, «external» magnetic dipole. Then $\bar{p}(y) = \text{const}$ is the equilibrium probability distribution for y in absence of the dipole x, and $\bar{p}(y/x)$ is the new equilibrium distribution in the presence of the dipole x. Resourse expense r is the energy of the interaction of two dipoles. Quantity $\beta = 1/T$, where T is the absolute temperature of the system. The higher the temperature T the less β, the weaker the interaction between y and x, and the less mutual information $I(X, Y)$.

If we have n independent external factors of the same type then we can write

$$\bar{p}(y/x) = \bar{p}(y/x^1, x^2, \ldots, x^n) = \exp[-n\beta r(x, y)]/Z \tag{1.52}$$

where $Z = \sum_y \exp[-n\beta r(x, y)]$.

Conditional entropy and mutual information between vectors x and y are:

$$H(Y/X) = n\beta R + \log \sum_y \exp[-n\beta r(x, y)] \tag{1.53}$$

$$I(X, Y) = H(Y) - n\beta R - \log \sum_y \exp[-n\beta r(x, y)]. \tag{1.54}$$

For instance, if we supply n identical external dipoles in the same states, then we can consider them as one «big magnet» consisting of n elements. Then mutual information between the state of this big magnet x and the state of dipole y is expressed by (1.54).

Now we can see what factors influence the conditional entropy and can diminish it. The more $n\beta$, the less the entropy $H(Y/X)$, the more the mutual information between the vector of parameters x and the state of the system y.

If the real probability distribution $p(y)$ deviates from the equilibrium one $\bar{p}(y/x)$, it aspires to return back to this equilibrium. In this case a spontaneous behaviour of the system can be directed to *decrease* the real entropy $H(Y)$. But the new objective function — mutual information $I(X,Y) = H(Y) - H(Y/X)$ — will increase in this process! If any random fluctuation or any external influence places the system in a state, where the entropy $H(Y)$ is more than $H(Y/X)$, the system will spontaneously return to the state with less entropy and more mutual information.

This process can be treated in many ways:

1. «Relaxation» — returning of the system to the equilibrium.
2. «Natural selection»: most «profitable» states (with the least resourse expense $r(x,y)$) have maximal probabilities $p(y/x)$ (i.e., they «survive»).
3. «Self-organization» — spontaneous decrease of the entropy.
4. «Adaptation» of the system y to the environment x.

The point of view exists that adaptation is a comparatively rare feature of matter inherent mainly to living beings. But the facts show that «fitness», accordance, and «harmony» are general properties of the world independent of whether this is the world of living beings or the world of physical bodies. The crystal of salt can not «survive» in pure water so wherever we find it there must be a dry environment. All things either adapt to their environment or are eliminated by it. All things that exist have been selected and adapted — more or less — to the environment. And, as a consequence, they carry the imprint of this environment, a smaller or greater amount of information about it.

Example: thermal interaction. If we place a body in «thermostat» the temperature of the body y approaches the temperature of the thermostat x. As we know in this case, the entropy of the system «body + thermostat» (at the microscopic level) increases. But, on the other hand, at the macroscopic level this process can be seen as the adaptation of the body to the conditions of the environment (= thermostat) in respect to such a macroscopic parameter as temperature. If we take an ensemble of such systems (to make it possible to calculate the probabilities), then we observe that the conditional entropy $H(Y/X)$ of the subsystem of the two macroscopic variables (x,y) (as well as conditional entropy $H(Y/X)$) will decrease in the process of relaxation (= adaptation!). At the initial moment of the process the distribution of the representative points (x,y) is random (see Fig. 1.6a), but towards the end of the process this distribution becomes more ordered: the temperature of the body y^k corresponds to the temperature of its «own» termostat x^k (Fig. 1.6b). As a result of this process the mutual information between the temperature of the body and the temperature of the environment increases, so that

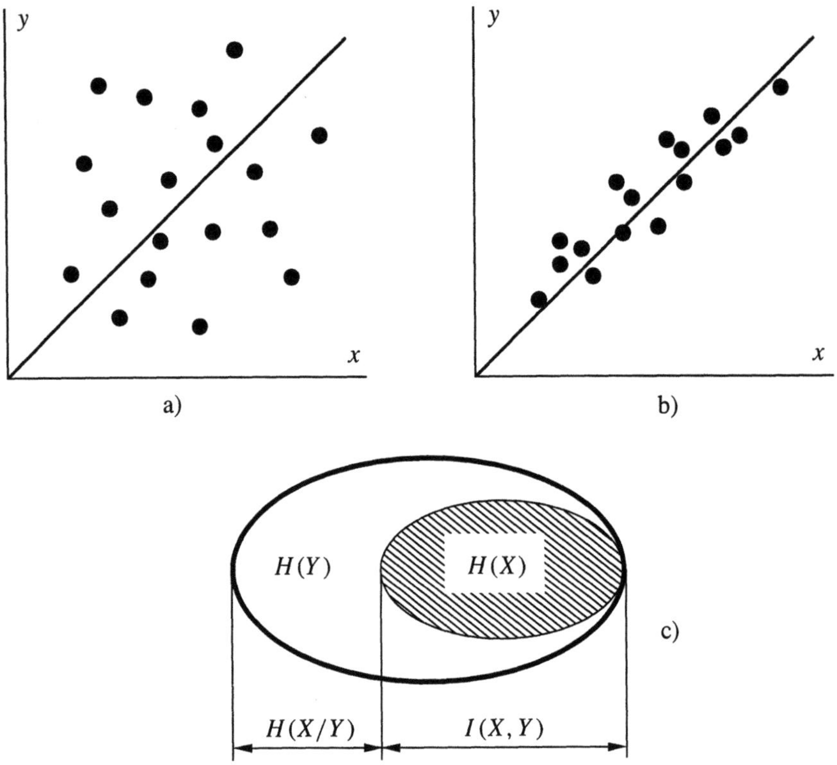

Fig. 1.6

when the process is completed we can make a judgement about the temperature of the environment by observing the temperature of the body.

When we place the thermometer into any media, wait until it stabilizes, and then read the information about the temperature of the media, we rarely think about such things as adaptation, selection, mutual information, etc. But in this case we have a primitive prototype of the more complicated and sophisticated processes of biological adaptation.

So we see that both processes of evolution — the growth of the unconditional entropy $H(Y)$ and the decrease of the conditional one $H(Y/X)$ can take place simultaneously. These two evolutionary processes are realized at different levels — microscopic and macroscopic. In physics the last process attracts not much attention because the number of the microscopic degrees of freedom is much more than the macroscopic ones, and the contribution of the last to the total entropy of the system is negligible. But in biology these processes — since they were more obvious and complicated — became the main object of investigation.

However, in physics the principle of adaptation of physical systems to the environment has been known (for a rather long time). This is the *Curie principle*

(1908). Adaptation is looked at in terms of symmetry. Symmetry of a physical system is determined not only by the inner properties of the system itself, but by the conditions of the external environment also. At the beginning of the century, Pierre Curie formulated the principle which connects the symmetry of the physical system with the symmetry of its environment: *the symmetry of the external influences* $S(X)$ *can not be more than the symmetry of the effects* $S(Y)$ *which are caused in the system by these influences:*

$$S(X) \leq S(Y). \tag{1.55}$$

One can see that the Curie principle is a more particular case of the principle of the information maximum. We can connect these principles by connecting the concept of symmetry with the concept of entropy: *the more symmetry, the more entropy.* Hence, the Curie principle can be written in terms of entropy:

$$H(X) \leq H(Y) \qquad \text{or} \qquad H(Y) - H(X) \geq 0 \tag{1.56}$$

where $H(X)$ is the entropy of the external influences;

$H(Y)$ — the entropy of the effects (= responses of the system).

This situation is pictured in Fig. 1.6c. In this case, $H(X) = I(X,Y)$, and $H(Y) - H(X) = H(Y) - I(X,Y) = H(Y/X)$ is the «noise» at the output of the informational channel. The less noise, the more mutual information, the more perfect adaptation. Hence, adaptation of the system to the environment looks like a decrease in the noise $H(Y/X)$ and, therefore, like the approximation $H(Y)$ to $H(X)$, i.e., like the approximation of the symmetry of the system to the symmetry of the environment.

Some examples:

1. A plant surrounded by homogeneous media must acquire the shape of a ball with a symmetry axis of infinite order. Some water-plants drifting in the water have exactly this shape (spherical chladoforae, for example). A deviation from this ideal form is often caused by the anisotropic action of gravity, flows, and other factors.

2. The shape of sea pebbles is not that of a ball but of a flat ellipsoid — because the pebbles exist in the non-isotropic media of the sea surf with the flat-parallel motion of the water.

3. The symmetry of an arising crystal depends on the symmetry or the dissymmetry of external media: the presence or the absence of gravity, the direction of the flow in a water solution, the proximity of the walls or of the bottom, etc.

4. Some asymmetric flowers acquire a symmetrical shape in the absence of constant gravity (for instance, if they are rotated during the period of flowering).

5. Symmetric forms of trees (in plane) are distorted by the wind blowing in a constant direction. The tree acquires the form of a «flag».

6. Usually the environment is not fully symmetric. Hence, adaptation to the environment demands of the system to lose its initial symmetry, to decrease the

entropy $H(Y/X)$. Living beings of a higher stage in evolution usually have a lesser degree of symmetry. This is an indicator of a more perfect adaptation to the conditions of the environment. On the other hand, regress in evolution is often accompanied by an increase of symmetry.

We may say that symmetry of the environment is reflected by the symmetry of the system which is formed in this environment. The system acquires and maintains only these elements of self-symmetry which coincide with the symmetry of the environment. Looking at the symmetry of the system we can form an opinion about the symmetry of the environment.

* * *

Information gives us an exact measure of progress, of development, and of organization. The process and the result of biological evolution, as well as of human activity — is the growth of information. The evolution of living beings, the history of mankind, the progress of culture, art, science, and technology — are at the same time processes of the information growth. The laws of development in these areas are the laws of information and can be obtained from the principle of the information maximum. It is the reason why information is the focus of our investigations.

In this chapter we've tried to reveal the reason why information plays such an important role: valuable information is the most general and adequate measure of adaptation of the system to the environment.

Chapter 2
Evolution and Creation

«... aufreibende Leistung, sei jedenfalls seine, des Führers und Veranstalters, in welchem der Wille Gehorsam, der Gehorsam Wille werde, dessen Person die Geburtsstätte beider sei, und der es also sehr schwer habe.»
«... Wer wird sich so quälen? Nennst du es Freiheit — diese Vergewaltigung deiner selbst?»

Thomas Mann, Mario und der Zauberer.

In this chapter we shall consider evolution as the emergence of new forms of adaptation, from the simplest ones to human creativity. We shall not touch upon real *mechanisms* of adaptation, which are very varied and complicated. We shall only consider their informational *principles*. There will be only a scheme, a logical framework of evolution. According to this scheme there are some important moments in this process:

- accumulation and use of energy;
- amplification of physical interaction and rise of informational one;
- rise of memory;
- rise of emotions;
- rise of perception;
- rise of symbols, and logical operations with symbols;
- rise of consciousness;
- rise of human creativity as a higher form of adaptation;
- rise of culture: language, art, science, and other cultural subsystems.

Some of these moments will be considered in greater detail below.

2.1 Life

Life can emerge as a result of the adaptation of matter to the comparatively narrow range of external conditions x^k. The main conditions are:

1. Presence of appropriate material resources — molecules of chemical elements like hydrogen, oxygen, carbon, sulphur, phosphor, etc.

2. Sufficient gravitation which concentrates the light molecules of some of these elements near the surface of the planet without letting them escape.

3. Water media which dissolves chemical compounds and provides the possibility for various combinations of elements.

4. Presence of sufficient energetical resources (sun radiation, electromagnetic discharges, thermal sources, etc.) which can overcome high potential thresholds and forms various types of bonds between the elements.

Life can be defined as the ability of the system for self-control and self-restoration on the basis of memory. Essential features of life are non-equilibrium; metabolism; and the accumulation and use of energy and other resources.

How are these features connected? Are these features random or necessary? Can life be based on any other features?

First of all we want to look at the concept of non-equilibrium. Any system can exist («survive») only near the state of stable equilibrium. Within the framework of our approach this is an equilibrium probability distribution $\bar{p}(y)$. Usually it is a state where any objective function of the system L becomes maximal. We shall name this state also the «normal» or «ideal» state. External forces usually force the system to deviate from this state, so that the *real* state $p(y)$ does not equal the ideal; and the objective function does not equal the maximum. Reactions of the system try to bring back the equilibrium, the «normal» state. Various systems are distinguished by the types of «forces», types of interactions which bring back and maintain the equilibrium.

But equilibrium is a relative concept. Let any external factor x interact with the system and it will cause a deviation from the equilibrium, so that the new state of the system $p(y/x)$ is non-equilibrium $p(y/x) \neq \bar{p}(y)$. But we can expand the framework of the system and include this factor in the system. Then the factor becomes an internal one, i.e., the external interaction disappears and the state of this enlarged system must be treated as a «normal» equilibrium.

For instance, the atomic nucleus is not in a state of equilibrium with respect to the electromagnetic interaction, but it is equilibrium in respect to the sum of electromagnetic and strong interactions. When we speak about the state of non-equilibrium of living systems we must take into account the relative character of any equilibrium and ask ourselves: in respect to which interactions do we consider equilibrium in this case?

In Nature we observe a hierarchical organization of systems. Each new level of organization is connected with the dominance of a new type of interaction. Life is no exception.

In physics we usually distinguish four types of interaction: strong, electromagnetic, weak, and gravitation. (This distinction is conditional and the list of interactions can be contracted as well as expanded.)

Many types of interaction are usually combined in one system, but as a rule there is one that dominates and determines the structure of the system and the type of equilibrium maintained. Strong interaction dominates in the organization and maintainance of nuclear structures, electromagnetic interaction — in crystals and chemical compounds, gravitation provides the equilibrium of planets and star systems — and so on. But what type of interaction dominates in the organization of living systems and provides a new type of equilibrium? This is the *informational* or *control* interaction: a new type of interaction that is absent from the physical world. But, like everything new, it is a combination of some old elements. These elements are:

1. *Translation* of any physical action in space or time.
2. *Amplification* of this action and creation of control forces which are added to the physical forces.

The first element — physical interaction — we have described (see Chapter 1) by means of the supplying item $\beta R(X, Y)$ to the objective function of the system. The simplest way to introduce the second element — amplification — is to multiply this item by the coefficient of amplification n. Thus we obtain the expression $n\beta R(X, Y)$ that must be supplied to the objective function of the system.

As follows from this expression, the effect of multiplier n is the same as the effect of increasing the number of stimuli (see Chapter 1): the more n the more «strong», «rigid», and «inertial» variable x is in comparing to variable y. For instance, if y is a magnetic dipole and x the totality of n magnetic dipoles, then x influences y, but y does almost not influence totality x. The interaction between x and y becomes asymmetrical, one-sided, directed from x to y. The more n, the more asymmetry of this interaction. The growth of the coefficient of amplification generates a similar effect of asymmetry.

This means that the increase of n decreases conditional entropy $H(Y/X)$ comparing with $H(Y)$, and increases the information from x to y; but it doesn't decrease $H(X/Y)$ comparing with $H(X)$ and doesn't increase the information from y to x! Translation of information becomes one-sided, asymmetrical. In our case it means that the system y reflects the state of the environment x, but the environment doesn't reflect the state of the system.

For instance, let x and y be two dipoles, the second dipole being supplied by an amplifier that can amplify the weak field of the first dipole n times; but dipole x has no such amplifier. It is easy to see that y can reproduce the state of dipole x, but dipole x cannot reproduce the state of y with the same exactness.

Only the consequent amplification transforms the physical interaction into an informational one and creates the basis for control processes. Translation and amplification as such exist, of course, in the physical world, but they are separated from each other. And only the combination of them creates a new effect.

This new type of interaction can be used, particularly, to maintain structures not in equilibrium. These structures are more interesting and promising from the point of view of the further evolution and adaptation. The reason for this is neither the «shortcomings» of the equilibrium as such, nor the «advantages» of the state of non-equilibrium as such. But it is easy to see that there are a lot more states not in equilibrium than states which are in equilibrium. Roughly speaking, equilibrium is a *point* in the space of states, while non-equilibrium is a *set* of states around this point. Hence, the probability of finding the most preferable, competitive, and adaptive combinations amongst those not in equilibrium is much higher. In other words, a non-equilibrium system contains, as a rule, much more useful information than one in equilibrium.

But these structures, not in equilibrium with the external environment, are rapidly destroyed in the interaction with this environment. And they have to be continuously restored to continue existing. This restoration — usually by means of full or partial reproduction — is dependent on two conditions:

1. Existence of memory — information about the values of the parameters to be restored.
2. Ability to use the resources of energy and materials.

The combination of these two conditions becomes the main feature of life.

If any state y was present in the past of the *surviving* system (i.e., this state *permits the system to survive*!), then there is a high probability that this state is favourable for survival in the future and so must be saved and reproduced. This is the reason why a familiar, habitual state becomes preferable for the *surviving* system. The proverb says «habit is a second nature». But we must add: «second and first!». All «normal», «preferable», and «optimal» values of the parameters of the living system are «habits».

Memory, informational interaction between past and present, and amplification provide control forces which maintain the equilibrium between the habitual past state of the system and the present one. We may say «equilibrium with its own past» (i.e., with the «habitual» states) is more important for the living system than equilibrium with the surrounding media, because it provides an advantage in survival.

2.2 Reflexion as a factor of evolution

The evolution of the non-living systems was connected with the dominance of various types of physical interaction and the arise of new types of equilibrium. Life is based on dominance of a new type of interaction — informational or control interaction. This interaction creates special control forces which permit the maintenance of these structures which are non-equilibrium in the physical sense, but may be considered as equilibrium ones in any new sense. The evolution of living systems can be seen as the emergence of new types of more and more complicated hierarchical structures. But all these structures are maintained *by means of the same type of interaction* — by the informational one. This interaction remains dominant in all stages of further evolution up to the emergence of human consciousness and human culture. This becomes possible because of a new property of this interaction, absent among the physical ones. This property we shall name «reflexion» or «reflexivity».

The term «reflexivity» in mathematics and logic means the ability of any operation or relation to return to itself. For instance, such relations as «to be equal» or «to be parallel» are reflexive: any quantity is equal to itself, any line is parallel to itself.

The term «reflexion» is also used in psychology and philosophy where it designates an ability of the human mental activity to turn to itself, an ability to control its own psychical life.

Finally, there exists the notion «reflex» introduced by R. Descartes. This refers to the ability of the living being «to reflect» on influences of the external environment, to answer these influences. The terms «reflex» and «conditioned reflex» are widely used in modern physiology and psychology.

All these meanings are connected with the notion «reflexion» which we use in this work. In general we can define a reflexion as the transformation of the *conditions* of the control («means») into the *aim* of the control («end»), as the transfer of the control from the *effect* to the *cause*. This looks like a «reverse motion» along the causal bonds from the end to the beginning.

We want to show that reflexion is one of the important principles of the evolutionary process. This principle provides the qualitative jump in the development of the system, the transition of life to a new level of organization. There are many «thresholds» and «jumps» on the way of evolution. One of them is the jump from the non-living matter to the living matter, when Nature had to build very large and complicated molecular structures (polypeptides and polynucleotides) to provide self-reproduction. The other well-known jump was the jump from the animal to man, the origin of human consciousness.

Many investigators felt special mechanisms had to exist to overcome such thresholds. *Hypercycle* proposed by M. Eigen (1951) is an example of such a mechanism. Reflexion is the general principle of the organization of such mechanisms.

Reflexion as a principle is inherent to the nature of the informational interaction as such: a control system can control itself, can be an object of self-control. There is a variety of mechanisms of control on different levels of organization, but the principle remains the same.

Reflexion allows a new control level to be «built up» on top of the old one and to supply compulsory processes to the spontaneous ones. But this supplement means, at the same time, transformation of compulsory into free, spontaneous behaviour — but *for the higher level of control*. Thus, reflexion means freer behaviour. Compulsion is transformed into self-compulsion.

One of the most important consequences of reflexion was the emergence of the hierarchical organization of the control processes in the living organism. A new level of hierarchy arises when any control process becomes the object (the «end») of another control process. It opens a possibility for the directed perfection and optimalization of the first process. One may say that all progress in the evolution of life is based on the hierarchical organization and optimalization of the control systems and processes by means of reflexion.

The result of such development was the emergence of *multi-level hierarchical control structures*. In these structures information is transmitted from the lower level to the higher one, where it is evaluated, selected, and processed. On the other hand, the higher level transmits to the lower one controlling influences, which determine the work of this level. See, also, Section 4.2. Free behaviour of each level is directed towards the maximum of the objective function L. But compulsory influences of the other levels give supplementary connections $R(X, Y)$ which change the form of the objective function of the considered level.

Usually we consider only two neighbouring levels of hierarchy — x and y. Here y is the level to be considered, x — any other, external level.

Now we want to look at a formal expression which describes reflexion. Let us suppose that an external connection is imposed on the state of the system and a new item $n\beta R(X, Y)$ is supplied to the old objective function. Then the new objective function L is:

$$L = \overline{L} - n\beta R(X, Y) = \max . \tag{2.1}$$

The last item in this expression describes exactly this reflexive process. It means that the variables n, β, R, x become *controlled* variables and the system aspires to change them to increase the objective function L.

To change R (or β) means to control resources, to transform them from the means into the end. We know well that some thing becomes the object of our attention and the aim of our activity when it becomes *deficient*. In the language of our theory this means that any Lagrange multiplier β (which characterizes a deficit of the resource R) becomes more than zero and the corresponding item βR is included into the objective function L. When resource R is sufficiently present we can use it, but we do not make it the end of our activity — we don't include it in our objective function ($\beta = 0$). It remains the means, but doesn't become the end. So man uses clean air and fresh water, but doesn't make them the object of special care until they become deficient.

We can say also that the resource R becomes a *valuable* variable — in the sense which was discussed in Chapter 1. We have noticed that the value emerges in the process of the exchange and can be defined as a relation of two exchanged quantities

$$V(y) = \frac{\Delta x}{\Delta y} \tag{2.2}$$

where V is a value of quantity y expressed in the units of quantity x. But any behaviour or activity can be considered an exchange of the resource R with the objective function L. Thus we can write for the value of the resource R:

$$V(R) = \frac{\Delta L}{\Delta R} = n\beta \tag{2.3}$$

We see that the value of the resource R is determined by the deficiency of this resource β.

Now let us consider how this concept — reflexion — manifests itself in the process of evolution.

The maintenance of a state of non-equilibrium with the surrounding media makes demands on the supplying resources. In the early stage of life these resources were sufficiently present in the surrounding media, so that living beings could use the resources from this media. But with the growth of the number of living beings the amount of resources began to decrease. The deficit β of the resource R arises, the item βR enters in the objective function L and a new aim emerges: to increase R and, as a result, to decrease β. This can be done in three ways:

1. *Economize* the resources. This can be done by means of the *optimal* choice of the probabilities p_{ij}.
2. *Store* the resource in any internal reservoir and *mobilize* it if necessary. This mobilization means a temporary increase of R and looks like *activation* or *excitation* — one of the main features of living beings.
3. *Produce* the necessary resources from *raw materials.* This permits R to increase on a constant basis.

Let us consider some examples of the last way which has occurred many times in the course of evolution. For instance, in the early stage of evolution living beings could use *hydrolysis* of «natural» phosphates from the surrounding media as a source of the energy. Soon there was a deficit of natural phosphates. Organisms began to produce phosphates using the process of *glycolysis* with glucose as a source of energy. Hence phosphates are transformed from the end into the means. Furthermore, when glucose became a deficit, organisms shifted to the production of glucose through the process of *photosynthesis* using CO_2 as a resource and the sunlight as a source of energy. Thus the process developed in the reversed direction from the end to the means.

Another example: in the first stage of life proteins were built on the basis of natural amino acids extracted from the surrounding media. But later on «free», natural amino acids became a deficit and living beings began to synthesize most of these amino acids inside the organism. The organisms were transformed from *heterotrophs* into *autotrophs*, and this transformation was one of the crucial moments of biological evolution and one more manifestation of the reflexive process.

In human history we also find many examples of similar transformations. First iron was used primarily as a raw natural material, but later when the need for iron increased, man switched to artificial production of iron. In a similar manner, hunting was replaced by cattle-breeding, collection of wild vegetables and cereals, agriculture, and so on.

2.3 Memory

We must distinguish two types of translations of physical action — in space and in time. The last one is usually called «memory». But in the exact sense, memory exists only under the condition of consequent amplification, which transforms the physical process into an informational one.

In this way, many physical systems are able to save a trace of some former influences. But only the amplification of this trace, and using it for control, transforms this ability into translation of the information in time, into *memory*.

We must add that amplification must be seen more broadly than usual. It is not only «energetical» amplification (which we find in electronics or in technical control systems), but also «material» amplification, such as copying, duplication, reproduction, manifolding, etc. Such types of amplification we can observe in the physical world (for instance — the growth of a crystal, when a new layer of atoms reproduces the old one). In living systems we find both types of amplification. It seems the first form of informational interaction was the translation in time and amplification by means of further reproduction — *memory*.

Mechanisms of memory are very various and complicated, but the mathematical model of memory is relatively simple. Let $y(t)$ be the present state of the system (at a moment t), $x^k(t^k)$ — a former state (in moment $t^k < t$). Let $\overline{p}(y)$ be an equilibrium probability distribution in the absence of memory (any interaction between x^k and y) and $H(Y)$ be the corresponding unconditional entropy. If the interaction exists, then it changes this probability distribution. If $r(x^k, y)$ is the resource expense, which is necessary to achieve the state y from x^k, we can find the new probability distribution

$$\overline{p}(y/x^k) = \exp[-\beta r(x^k, y)]/Z \tag{2.4}$$

and conditional entropy

$$H(Y/X^k) = \beta R(X^k, Y) + \log x \sum_{x,y} \exp[-\beta r(x^k, y)]. \tag{2.5}$$

The average mutual information which is contained in the present state y about the past one x^k:

$$I(Y, X^k) = H(Y) - \beta R(X^k, Y) - \log \sum_{x,y} \exp[-\beta r(x^k, y)] \tag{2.6}$$

is a quantitative measure of memory.

Here r can be time, energy, etc.; β depends on the deficit of time, energy, etc. The less β, the more $H(Y/X^k)$, the less information $I(Y, X^k)$: the system «forgets» its past.

The distribution $\overline{p}(y/x^k)$ is a compulsory one, it is conditioned by the connection of the present state y with the former one x^k. But, on the other hand, if we include this connection in the structure of the system, we can see this compulsion as self-compulsion and this distribution as a new equilibrium. Memory is, of course, a typical example of self-compulsion. It maintains an equilibrium of the present state of the system with the past one. If any state of the system $p(y)$ deviates from $\overline{p}(y/x^k)$, the memory provides the control forces, which force the system back into this state.

For instance, let us look at a «magnetic dipole with a memory». The memory means that the past state of dipole x^k becomes the equivalent of the existence of some external dipole. This dipole creates a magnetic field which influences the present state of the dipole y. Removal of the dipole x^k from y in time (as well as removal in space) attenuates the effect of memory. A rising temperature of the dipole (and, therefore, a decrease of β) leads to the same effect — «forgetfulness».

But memory becomes the means to overcome this effect. Usually the memory can store $n > 1$ former states of the dipole x^k. Then each new step in time increases the number of «external dipoles». The «mean field» of these dipoles becomes more and more strong — the memory works like an «amplifier». Conditional probability distribution $\overline{p}(y/x^1, x^2, \ldots, x^n)$ becomes more and more «acute», i.e., high probabilities increase and low probabilities decrease. This looks like a *selection* of the most informative states y of the dipole. The process becomes unstable, proceeds with positive feedback, and aspires to the ideal state. In this process conditional entropy $H(Y/X)$ decreases and the average mutual information between the past and the present $I(Y, X)$ aspires to a maximum. Thus, the memory is a *cooperative* process.

An important feature of memory is *plasticity*. The depth of memory cannot be infinite. The volume of the former experience — the set of states $\{x^1, x^2, x^3, \ldots, x^n\}$ that influences the present state is always restricted ($n \neq \infty$). When a new (present) state y is included into the set, an old state x^i must be excluded from it, so the content of the memory can slowly change. The system *forgets* the old state.

This feature of memory is a basis for such properties of the living systems as *plasticity*, adaptiveness, ability to learn and re-learn, formation of new habits, etc. Because of this feature a system, which finds itself in any new state for some time, can memorize this state and maintain it for some time.

Objective probabilities $p(x), p(y), p(x, y)$ are reflected by the memory in the form of subjective probabilities. They become internal, controlled variables and can be changed in the desired direction, which then provides an increase of the objective function L. In other words, the ability of spontaneous idealization emerges.

Memory causes certain properties of the living system, like threshold, the overcoming of threshold, excitation, etc. to emerge.

Let us look at some of the stages in the development of memory.

The prototype of memory, which is able to maintain a non-equilibrium state and defend it from the influences of the external media, are the various types of *inertia*. If we force a brick into a position of non-equilibrium above the Earth's surface and then release it, it doesn't shift to the state of equilibrium immediately.

The force of inertia wants to preserve the initial state, and only gradually the position of the brick changes. The more the mass of the brick (the number of «elements» of the brick n), the more the force of inertia. In various cases we see mechanical, thermal, chemical, and other forms of inertia. Usually the higher the number of elements of the system n, the more the inertia, the more the time constant τ, the smaller the rate of evolution of the system towards the equilibrium state, under the influence of the external environment. Hence, the growth of the number of elements n is a form of response of the system; and inertia is a form of control which permits the system to save its original state. Let us look at some examples.

We want to start with the old question: why are atoms so small? To answer this we have to reverse the question: why are our bodies so large in comparison with atoms? It seems that the large size and the inertia of our organs are necessary conditions for the stable performance of their functions.

For instance, why do proteins have such a large molecular weight? Only a few parts of a protein molecule have fermentative activity. And what is the function of the other, «passive» parts? Why were they not eliminated in the course of evolution in spite of the fact that they demand a lot of energy, materials, and other resources? One of the possible explanations is: the inertia of the protein molecule is a necessary condition for its normal stable work. It provides a stable conformation of the protein molecule and a stable configuration of the active centers under the conditions of continuous disturbances from the molecules of external media. Similarly, the machine must have a hard massive foundation to provide the necessary exactness and stability of work.

A membrane is the other useful «invention» of the evolution, based on the property of inertia. A membrane, first of all, divides the media into two parts — internal and external — and separates the internal part from the external one. It increases the resistance against various flows from the external environment into the internal space. Hence, time constants for the corresponding transitive processes — chemical, thermal, mechanical, etc. — increase. Thanks to the membrane many chemical elements and compounds in the internal media preserve non-equilibrium concentrations; and the fluctuations of the parameters in the internal media are much smaller than those in the external one.

This «invention» has been repeated in various forms in other stages of evolution. The shell of the snail and the tortoise, the hair of the wolf, the clothes of man — are all a demostration of the same principle: isolation from the external media by means of any type of cover which possesses some kind of «inertia». The simplest example is the protein cover of a virus which defends an internal part — the informational RNA molecule — from the external media.

Note, inertia in living beings is usually not a passive, but an active response: it demands a resource expense to create and maintain the structure of the cover. Moreover, it is a non-economical response, because it is not flexible and adaptible enough. This response hardly changes over time, it cannot rapidly be «switched off» whenever the external disturbance disappears. It is almost an «uncontrolled» response. The next, more perfect form of response — the controlled response — had to emerge.

The simplest form of controlled response is «control based on deviation». The system of control is switched on when the state of the system $p(y)$ deviates from the normal, habitual state $\bar{p}(y)$, under the influence of any external disturbance. This can economize resources, as they are not spent when deviation is absent. But this type of control is not perfect in essence: deviation must *exist* to provide the normal functioning of the control system, i.e., in principle it cannot be completely eliminated. It had to cause the emergence of the next, more perfect form of control.

The next stage of control development is «control based on disturbance». The system doesn't wait for any external disturbance to cause a deviation of the internal state, but works «prophylactically»: it switches on as soon as the disturbance emerges. In other words the control is transferred from the effect (deviation) to the cause (disturbance), from the internal parameters y to the external ones x. External parameters becomes the end of control and the system aspires to stabilize these parameters and prevent disturbances — deviations from the normal values.

This type of response demands the system's *perception* of the external media and its *anticipation* of the effects of external events («stimuli») x on the internal variables y. The memory becomes the basis of this anticipation. And the matrix of probabilities $p(x, y)$, stored in the memory, becomes the model for the connections between x and y. Conditioned reflex is a well-known example of this connection. Uneveness of the probability distribution $p(x, y)$ means that there are «habitual» (high probability) and «unhabitual» (low probability) combinations (x, y).

2.4 Response of the system to a new stimulus

First of all, we must define the terms «new» and «old». In the language of the information theory «new» means «of low probability». «Old» or «habitual», on the contrary, means «of high probability». Note, in accordance with Kolmogorov's approach, we can replace probability also by «frequency», «concentration», «relative part», etc.

With the rise of memory the *habit* appears. And the system obtains an ability to estimate the probabilities (frequencies) of the events and to distinguish between the *habitual* («old») and *unhabitual* («new») events.

Now we want to draw your attention to the fact that «new» and «old» are relative concepts: if any new situation emerges and continues to exist, then its probability increases and the situation becomes old, habitual. «Growing old» equals «growing of probability».

On the other hand, the memory of the system is usually limited, so that an old event can be forgotten, replaced by a new one. In this case the probability of the first one will decrease, old becomes new: «the new is the well-forgotten old».

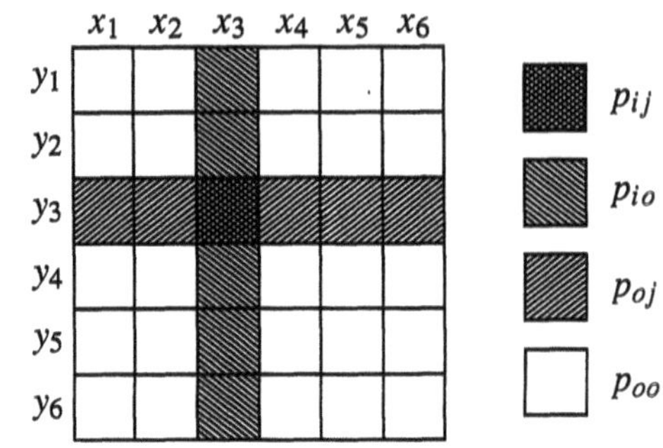

Fig. 2.1

Let us consider how $I(X, Y)$ changes when a stimulus x_i acts, the system answers with the fixed response y_j, and as a result the probability $p(x_i, y_j)$ increases from 0 to 1. We shall designate $p(x_i) = p_i$, $p(y_j) = p_j$, and $p(x_i, y_j) = p_{ij}$.

At first, we shall take a case where a deficit of resources is absent ($\beta = 0$) and the last item in (1.17) can be omitted. We may also neglect the condition of normalization and, correspondingly, the Lagrange multiplier α. Let us pick up the item with the probability p_{ij} and the sum of all other items designated as S:

$$I(X, Y) = p_{ij} \log \frac{p_{ij}}{p_i p_j} + S = p_{ij} \log \frac{p_{ij}}{(p_{ij} + p_{i0})(p_{ij} + p_{0j})} + S. \tag{2.7}$$

Here p_{i0} is the sum of all «other» probabilities in p_i (except p_{ij}): $p_{i0} = \sum_k p_{ik}$, $k \neq j$. So is $p_{0j} = \sum_l p_{lj}$, $l \neq i$. (See Fig. 2.1.)

Let us consider the qualitative picture of the dependence of mutual information $I(X, Y)$ in respect to probability p_{ij}. This dependence in general passes through three stages (Fig. 2.2a):

I. *Attenuation.* When p_{ij} is zero (or near zero) the denominator in the logarithm $p_i p_j$ in general is not zero. Hence the logarithm is negative and the information decreases with the growth of p_{ij}. As soon as it contradicts the aim of the system, it causes a negative attitude of the system towards the stimulus. The system attempts to reject such a stimulus, to attenuate, to inhibit, or to destroy it. In this stage only a physical interaction between the system and the environment exists; an «informational contact», informational interaction, is still absent.

II. *Amplification.* As soon as p_{ij} grows more quickly than $p_i p_j$, at a certain moment the logarithm can become positive. In this case the growth of p_{ij} will lead to the increase of information $I(X, Y)$. The attitude of the system to the stimulus becomes positive. At this stage the system aspires to attract and to assimilate the stimulus, to amplify, and to prolong it. The second stage is a stage of the informational interaction between the system and the environment.

III. When p_{ij} approximates 1 then mutual information $I(X, Y)$ goes to zero. As soon as information is a positive quantity, in the third stage information decreases and the attitude of the system to the stimulus becomes negative again.

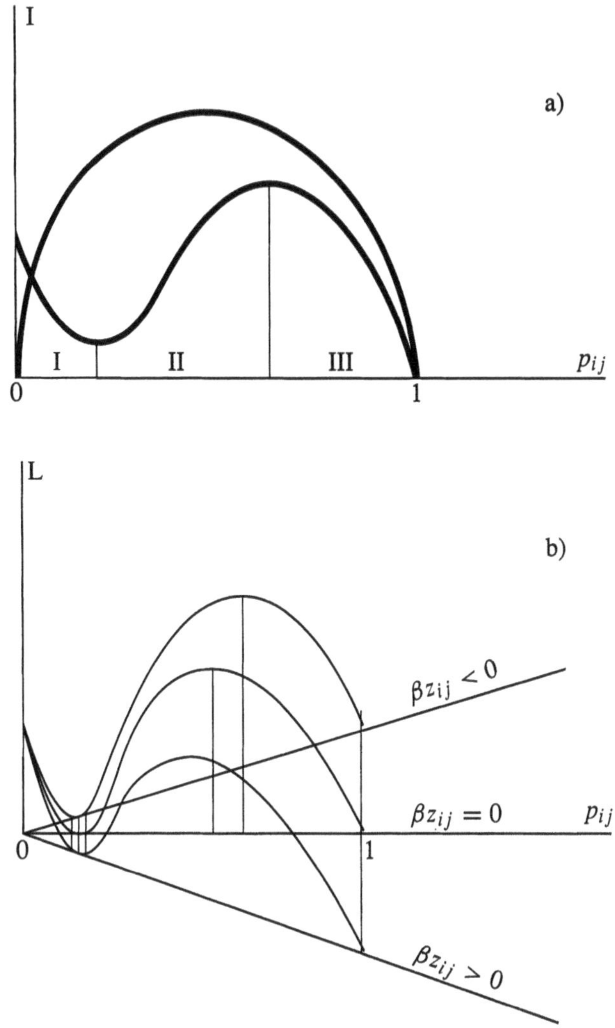

Fig. 2.2

This stage we shall call *«stagnation», «depression», «fatigue»*. The only means to break off this stage is to switch to any «other», «contrast» stimulus.

Simonov (1962) was the first who investigated these three stages systematically. He generalized a large amount of facts and defined these stages in the language of physiology as «initial or preventive inhibition», «activation or excitation» and «supermarginal inhibition».

The whole picture is displayed in Fig. 2.2a. Compare it with Fig. 1.4a which contains only two stages and one point of extremum. A new moment is the first negative stage which is absent in Fig. 1.4 and *three* points of extremum. We see here two points of maximum $I(X, Y)$ and one point of minimum between them. This point of minimum plays the role of threshold which divides the areas of attraction of the two maxima. This new moment is conditioned by the presence of the second item — conditional entropy $H(X/Y)$ — in the objective function.

It is useful also to describe these three stages in terms of entropy and organization. Information is a measure of organization, particularly, a measure of connection, correlation between variables x and y. Hence, we can say that the system passes through the stages:

I. *Disorganization.* A decrease in information in the first stage, near the threshold, means the breakdown of the old, habitual connections, the destruction of the old structure of the system. This breakdown of connections means that the number of degrees of freedom and entropy (variety) of responses $H(Y)$ (as well as entropy of errors $H(Y/X)$) sharply increase. The «area of activity» of the system (area of high probabilities in the probability matrix) sharply expands.

Near the threshold we can observe some interesting effects, many of them are well known. In physics we see the growth of instability, entropy, and fluctuations near the point of the phase transition. In psychology we observe similar effects under the sharp change of the way of life, «re-learning». They look like «activation», «expansion», and «search». Similar effects accompany also a reorganization of any social system («perestroika» in Russia, etc.).

II. *Organization.* The probability of a new combination repeated, increases and overcomes the threshold. New correlations (connections) between the variables arise, the number of degrees of freedom, entropy of responses, and entropy of errors $H(Y/X)$ decrease. Mutual information $I(X, Y)$ increases. It is a stage of concentration of the system on a new «area of activity», a stage of selection and idealization of the most optimal responses, traits, and patterns of behaviour.

III. *Stagnation.* The new combination (x_i, y_j) little by little becomes an old one, the probability p_{ij} grows and grows, approximating 1, entropy $H(Y)$ and mutual information $I(X, Y)$ diminish. In this stage the area of the system activity is contracted to the single position on the probability matrix. This situation can become dangerous for the system: the greater the probability of this combination, the less the threshold for any other unexpected and contrast combination, and the more destructive for the system any new stimulus will be.

Now let us consider how additional conditions, such as resource deficiency, influence the form of the graph $I(p_{ij})$. We must return to the whole expression (1.17) and present it in the form

$$L = p_{ij}(\log \frac{p_{ij}}{p_i p_j} - \alpha - \beta r_{ij}) + S \tag{2.8}$$

where S is the sum of items without p_{ij}.

In Fig. 2.2b graphs $L(p_{ij})$ with various values βr_{ij} are shown. One can see that the higher βr_{ij}, the higher the threshold, the lower the optimum. When βr_{ij} is large enough, the threshold and the optimum merge and there remains only one point of maximum — $p_{ij} = 0$. This means that if the resource deficiency β and the resource expense in the state (x_i, y_j) are finite but too large, the probability of this state can only be zero.

In the previous chapter we have received a probability distribution (1.14) which provides the entropy maximum. Now we want to consider the similar distribution which provides the maximum of the mutual information $I(X, Y)$. It will be a bi-dimensional distribution in the axes x and y. But the simple technique which we used in the previous chapter to search the optimal distribution $p(y)$ — to equalize the partial derivative of L to zero as a condition for the maximum — does not work in the case considered. The reason is that the maximum of $I(X, Y)$ can be achieved on the boundary of the permissible area of values of p_{ij} — in the point where some probabilities $p_{ij} = 0$. In this point the partial derivative is not zero, in spite of the extremum that exists. For this case we can use a numerical procedure which can be carried out by a computer.

We must distinguish between two different cases:

1. Both the stimulus x_i and the response y_j are fixed, and have compulsory values so that probability $p_{ij} = p(x_i, y_j)$ increases with time. It can be both external compulsion, or internal self-compulsion. Learning is an example of this situation.

2. The stimulus x_i is given, fixed, but the response y_j is free, so that the subject can choose an optimal value of y_j. If the subject prefers the response y_j and rejects another response y_0, then the probability $p(x_i, y_j)$ increases and the probability $p(x_i, y_0)$ decreases. Free, spontaneous behaviour is an example of such a situation.

In the process of learning the habits, connections between the variables are formed. It is expressed in the formation of the matrix of probabilities.

In behaviour these habits, or connections, begin to «work». They compel the system to choose the responses which are «optimal», i.e., maintain and reinforce the existing habits and connections; and to reject the responses (and the stimuli) which threaten to destroy old habits and connections.

But partial destruction and reorganization of the connections is inevitable in the process of development of the system.

If the response is free (variable y is not fixed), the subject can choose a value of y and, therefore, either increase or decrease probability p_{ij}. The threshold is a point of instability and near the threshold there are two ways in which probability p_{ij} can behave: if the initial value of p_{ij} is less than the threshold, then p_{ij} goes to zero, if more — then p_{ij} goes to the optimal value $\overline{p}_{ij} > 0$. This process can be called «idealization». It looks like this: if a system receives any stimulus x_i, it selects as its response a value y_j, so that the probability p_{ij} lies above the threshold, and rejects values y_0 which have probabilities p_{i0} under the threshold. As a result, the first probability increases; the last ones decrease. During this process the large probabilities increase, the small probabilities decrease, and the whole picture aspires to an optimal, «ideal» distribution. In particular, this distribution can look like a diagonal matrix where probabilities are concentrated on the diagonal and evenly distributed along it. In this case the conditional entropy $H(X/Y)$ is zero, the unconditional entropy $H(X) = H(Y)$ is maximal, and the mutual information $I(X,Y)$ is maximal also. On the curve $I(p_{ij})$ the first stage at first diminishes and then ceases to exist during the process of diagonalization, and the curve $I(p_{ij})$ in Fig. 2.2a becomes similar to the curve $H(p_j)$ in Fig. 1.4a.

In a more general case, «ideal» distribution looks like a matrix of conditional probabilities $p(x_i/y_j)$ which contains a 1 in the single position of each line; the other positions are occupied by zeroes.

2.5 Conditions of the informational interaction. Threshold

Now we want to find the conditions to overcome the threshold through a new combination (x_i, y_j). They are at the same time conditions of informational contact and informational interaction between the system and its environment — particularly conditions of perception of the object by the subject. There are many objects around the subject, and interaction with all of them at the same time is impossible. Most of these objects find themselves under the threshold of interaction and only a few of them are in informational contact with the subject. Thus the conditions for this contact have to be found.

As is clear from Fig. 2.2, the threshold is overcome when a partial derivative $\partial L/\partial p_{ij}$ is equal to zero. Hence, we can write the condition of the threshold:

$$\partial L/\partial p_{ij} = \log \frac{p_{ij}}{p_i p_j} - 1 - \alpha - \beta r_{ij} = 0. \tag{2.9}$$

Transforming expression (2.9) we obtain for p_{ij}:

$$p_{ij} = p_i p_j \exp(\alpha + 1 + \beta r_{ij}) = p_i p_j \exp(\beta r_{ij}) \exp(\alpha + 1). \tag{2.10}$$

The multiplier $\exp(\alpha + 1)$ can be derived from the condition of normalization:

$$\sum_{ij} p_{ij} = \sum_{ij} p_i p_j \exp(\beta r_{ij}) \exp(\alpha + 1) = 1. \tag{2.11}$$

Therefore

$$\exp(\alpha + 1) = 1/\sum_{ij} p_i p_j \exp(\beta r_{ij}) = 1/Z \tag{2.12}$$

where Z is the so-called «statistical sum». But this quantity has no simple and obvious physical sense and it is easier for us to express it in terms of the average information I and the average resource expense R. Averaging expression (2.9) over all i and j we obtain:

$$\sum_{ij} p_{ij} \left(\log \frac{p_{ij}}{p_i p_j} - 1 - \alpha - \beta r_{ij} \right) = I - 1 - \alpha - \beta R = 0 \tag{2.13}$$

or

$$-1-\alpha=\beta R-I. \tag{2.14}$$

Substituting (2.14) in (2.9) we obtain:

$$\partial L/\partial p_{ij}=L_{ij}-L=\log\frac{p_{ij}}{p_i p_j}-I-\beta(r_{ij}-R)=0. \tag{2.15}$$

Now we want to express p_i, p_j through the probabilities of the «other» stimuli and responses:

$$p_i=p_{ij}+p_{i0}, \qquad p_j=p_{ij}+p_{0j}. \tag{2.16}$$

where p_{i0} is the sum of all «other» probabilities in p_i except p_{ij}; so is p_{0j} (see also Fig. 2.1).

Now we can rewrite (2.15) under the form:

$$\begin{aligned}\partial L/\partial p_{ij}&=\log\frac{p_{ij}}{(p_{ij}+p_{i0})(p_{ij}+p_{0j})}-I-\beta(r_{ij}-R)\\&=\log\frac{p_{ij}}{p_{ij}(p_{ij}+p_{i0}+p_{0j}+p_{i0}p_{0j}/p_{ij})}-I-\beta(r_{ij}-R)=0.\end{aligned} \tag{2.17}$$

Taking into account that $p_{ij}+p_{i0}+p_{0j}=1-p_{00}$ (see Fig. 2.1) and contracting p_{ij} we can rewrite (2.17) as follows:

$$-\log(1-p_{00}+p_{i0}p_{0j}/p_{ij})-I-\beta(r_{ij}-R)=0. \tag{2.18}$$

Or

$$1-p_{00}+p_{i0}p_{0j}/p_{ij}=\exp[-I-\beta(r_{ij}-R)]=0. \tag{2.19}$$

Or

$$p_{ij}-\frac{p_{i0}p_{0j}}{p_{00}-1+\exp[-I-\beta(r_{ij}-R)]}=0. \tag{2.20}$$

It is the most appropriate expression for the condition of overcoming the threshold.

There are many theories of threshold, but they consider only part of the factors that determine the threshold. Our approach to the problem (which can be named the «informational theory of threshold») permits covering most of these factors. These factors are contained in expression (2.20), but we must give them a physical and psychological interpretation:

a. *Sensibilization.* Let us assume that the combination (x_i, y_j) already has been exposed in the previous experience. The greater the exposition, the higher the probability p_{ij}; the greater the left part of expression (2.20), the lower the threshold for a new exposition of the combination (x_i, y_j). That is, the previous exposition of the stimulus facilitates the further perception of it. This effect is known as «sensibilization», «tuning» of the system to the stimulus, or «dominance» of the stimulus. It is the reason why a repeated, regular, rhythmical stimulus is perceived more easily than an irregular one.

b. *Concretization.* Supplying to the stimulus any new feature, any new component x^k of vector x leads to an increase in the information in the stimulus. Correspondingly, the threshold for the stimulus decreases. This is a very important result: the more exact, complicated, concrete the stimulus, the more features (components of vector) n we use, the lower the threshold, the easier the perception of the stimulus.

c. *«Wrong habits».* If probabilities p_{i0}, p_{0j} are not zero, it means that in the former experience of the subject there were «wrong habits», i.e., when the subject has answered «this» stimulus x_i with any «other» response y_0, or with «this» response y_j any «other» stimulus x_0. These habits lead to the increase of the threshold for the answer with «this» response y_j to «this» stimulus x_i. In other words, a habit of «incorrect» responses to a stimulus prevents the «correct» response to it and increases the threshold for this response — this is a well-known fact.

d. *Contrast.* If an «other» stimulus x_0 is accompanied by an «other» (i.e., «correct», «adequate») response y_0 in a former experience the picture changes. The more normal an «other» adequate response y_0 is to an «other» stimulus x_0, the more the probability p_{00}, the less the threshold for «this» response y_j on «this» stimulus x_i will be. A well-known example is the adaptation to darkness: the longer the time of adaptation, the lower the threshold for perception of light. This regularity forms the basis for such an aesthetic device as «contrast»: «this» stimulus must have a large contrast («other») stimulus in the background (temporal or spatial) to be easily perceived.

e. *Informativeness of the background I.* The higher the average informativeness of the background I of the stimulus, the higher the threshold. It is a well-known fact that a bright light, a loud sound, an interesting conversation, and other informative stimuli in the background distract the attention from a given stimulus and increase its threshold of perception. In daytime we cannot see stars nor hear quiet sounds which are easily perceived at night.

f. *Expense of the resource r_{ij}.* If the expense of the resource r_{ij} increases, then the threshold of the response y_j, naturally, increases also. So an extremely high resource expense («potential barrier») can completely prevent the response.

g. *Deficit of the resources β.* First of all we must distinguish betwen two types of responses:

1. *Active response* («excitement», «strain») when the resource expense r_{ij} in the state (x_i, y_j) is higher than the average level, $r_{ij} > R$. This type of response is the most typical one for living beings. Such a response demands an additional expense of energy and is in need of a special source of energy. Using such sources of energy is one of the most significant features of life. Active response neglects the expense of energy for the sake of a more perfect adaptation.

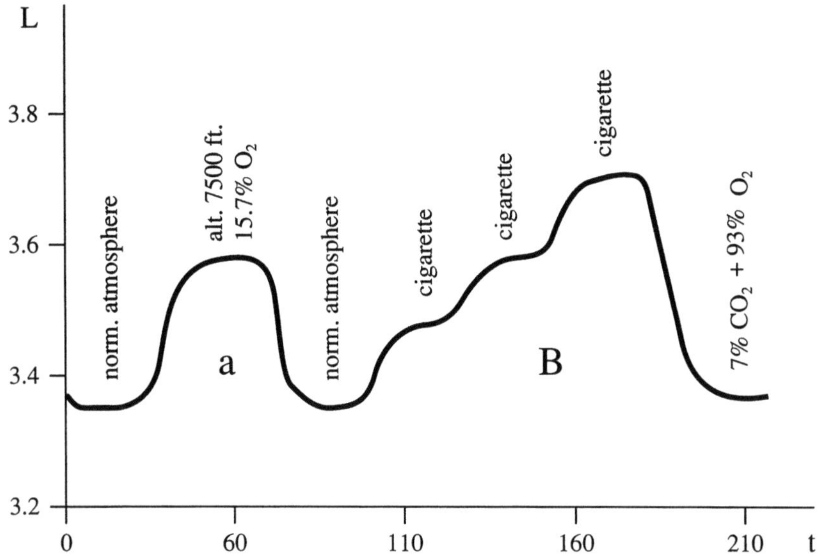

Fig. 2.3

2. *Passive response* («rest», «relaxation») when $r_{ij} < R$. This response is not typical for living beings. Maybe this is the reason why a passive response is usually not considered a response, but an absence of any response.

The threshold of the active response increases when the deficit of the resources β increases. Such factors of energy deficit as hunger, exhaustion, fatigue, hypoxia, poisoning, illness, and old age have to increase the threshold of the active response. There are many facts confirming this conclusion.

On the other hand, decrease of β, mobilization of the resources, decreases the threshold.

For the passive responses the dependence of the threshold on the resource deficiency is exactly opposite.

These conclusions are confirmed by numerous and well-known facts. Some of them are presented in Figs. 2.3 and 2.4. They demonstrate a dependence of the threshold of the visual perception of some factors: hypoxia, nicotine poisoning, hunger, and time of adaptation to darkness. In this case the stimulus x_i is light, the response y_j — the sensation of light (active response!).

Fig. 2.3 (McFarland, 1946): Deficit of resources caused by hypoxia (a) and by nicotine poisoning (b). Axis L is the logarithm of the threshold. Axis t — time (min).

Fig. 2.4 (Kravkov, 1950): Deficit of resources caused by hunger. The three curves picture different degrees of exhaustion: *1* – 0 days of starvation, *2* – 44 days of starvation, and *3* – 14 days after starvation. On the axis S — sensibility,

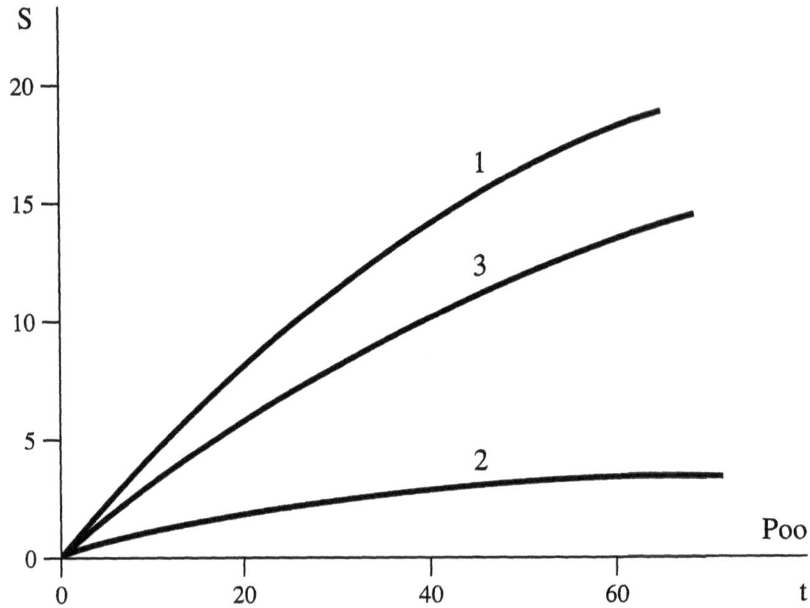

Fig. 2.4

the quantity which is inverse to the threshold. On the axis t — time of adaptation (min). Probability p_{00} characterizes adaptation to darkness. This quantity grows with time of adaptation.

As far as the passive responses are concerned we don't find such well-known facts, because these responses don't get much attention and often are not considered to be a response at all. But still a typical passive response like sleep exists. In this case p_{ij} is the probability of sleep response y_j on the sleep stimulus x_i. Usually this stimulus is darkness, night time. The subject can answer this stimulus using two types of responses: «adequate» y_j (to sleep) and «inadequate» y_0 (to be awake, to work at night, etc.). On the other hand, there are also two types of responses to the other stimulus x_0 (light, daytime): «adequate» y_0 (to be awake) and «inadequate» y_j (to sleep during the day). Note, «to be awake» is an active response $(r_{ij} > R)$; «to sleep» — a passive one $(r_{ij} < R)$.

Now we can derive from expression (2.20) what factors decrease the threshold and promote sleep:

a. *Resource deficiency* β. As soon as $(r_{ij}-R)$ is negative: the more β, the lower the threshold. Tiredness, hypoxia, illness, and other factors of resource deficiency favour sleep.

b. *Resource expense in the sleep* r_{ij}. More comfortable conditions promote sleep; harder, uncomfortable conditions (high level of r_{ij}) — prevent it.

c. *Resource expense in the other, alternative activity* y_0. The more r_{00} (which is included in the average value R), the harder this activity, the lower the threshold of sleep. For instance, the reading of a difficult text (especially at night) can provoke sleep.

d. *Informativeness* I_{00} *of the other, alternative activity* y_0 (which is included in the average value I). The more I the higher the threshold. So, interesting books, conversations, thoughts can prevent sleep. On the other hand, «boredom» (low informativeness I_{00}) of the other activity y_0 (a boring lecture, boring book, etc.) can lower the threshold and promote sleep.

e. *«Wrong habits»* — to be awake at night or to sleep during the day increase the mixed probabilities p_{i0}, p_{0j} and increase the threshold for sleep.

The problem of the threshold is very important to understand the perception of any stimulus. It is the problem of «informational contact» between the system («subject») and its environment («object»). Any perception and cognition of an object is impossible without this contact. Independent of the physical and physiological mechanisms realizing this contact, the properties and regularities of this contact remain the same because they are based on the general principles of information processing. We could look at any other informational channel, for instance, memory — and find the same regularities. Memory can be seen as a «sense» allowing us to perceive objects, removed in time, like the sight is a sense allowing us to perceive objects removed in space. Hence, remembrance is a kind of perception and the conditions of overcoming the threshold («informational contact with the past») must be the same.

If any «extrasensory» informational interaction exists, then the conditions of the informational contact and overcoming the threshold must be valid for this exotic «sixth sense» as well as for the usual perception channels — sight, hearing, memory, etc.

This problem is also directly connected with the problem of creation. Insight, guessing, the solving of a creative task, can be considered to be the overcoming of some threshold, the «informational contact» with the solution.

2.6 Emotions

Reflexion obtains a new direction after the emergence of the psyche. This is the direction towards the internal world of the subject. The psyche is the means of control of our behaviour. But the means becomes the end: the psyche becomes an object of control for the higher control level. This new control level is made up of the emotions.

The problem of emotions is one of the most difficult in psychology. What is the function of emotions in behaviour? What factors determine the value and the sign (positive or negative) of emotion? How do emotions relate to the other determinants of behaviour?

One of the most interesting approaches to the problem of emotions — «biological theory of emotions» — is that of the famous Russian physiologist P.K. Anokhin (1978). He saw emotion as the criterion which makes it clear to an organism whether he moves towards a goal or away from it. In the first case emotion is positive («pleasure»), in the second one — negative («pain»). Thus pain and pleasure have an adaptive significance and help the organism to achieve its goals and to survive.

If we have a mathematical expression for a goal — an objective function L — we can bring down emotion to the following expression:

$$E = dL/dt = \partial L/\partial q \cdot dq/dt = F_q \cdot \dot{q} \tag{2.21}$$

where q is any variable on which emotion depends.

We have divided this expression into two multipliers. The first one F_q is a controlling «force» which acts on variable q. In physiology and psychology this force is known under the names «need», «drive», «motivation», etc. The second multiplier is the rate of change of variable q. This can be interpreted psychologically as the rate of satisfaction of the need. An emotion is positive when the directions of the drive and of the rate of satisfaction coincide; if this is not so the emotion is negative. The value of the emotion is proportional to the value of the drive and to the rate of its satisfaction. These general conclusions are well known and are supported by numerous facts.

Now let us look at the main functions of emotions.

We have seen that the change of such variables as probability p_{ij} can be useful (increase of the objective function L) as well as harmful (decrease of L). Hence, in the complicated media the choice of the response on the given stimulus is not fixed once and for all — the response must be chosen taking into account many surrounding conditions: p_{i0}, p_{0j}, p_{00}, I, β, r_{ij}, R, etc. The system must have a general criterion which can integrate all these conditions and make a decision: to increase this variable or to decrease it. Emotion becomes exactly this criterion. This is a general function of emotion. But there are also particular functions which depend on the type of variable to be changed. Let us look at this problem in detail.

We can write out the expression for the emotion like this:

$$E = \frac{dL}{dt} = \sum_{ij} \frac{\partial L}{\partial p_{ij}} \frac{dp_{ij}}{dt} - \frac{\partial L}{\partial R} \frac{dR}{dt}. \tag{2.22}$$

Partial derivatives $F_p = \partial L/\partial p_{ij}$ and $F_R = \partial L/\partial R$ we can treat as controlling forces that provide the change of the corresponding variables p_{ij} and R in the desired direction. As soon as probability p_{ij} is dependent on the response y_j, we can say that there are three main variables which the system can change — p_{ij}, y, and R. And, correspondingly, three main functions of emotions:

1. Learning: to judge the probability p_{ij} useful or harmful and, correspondingly, to increase the first and to decrease the second one.

2. Behavior: to judge the response y useful or harmful and, correspondingly, to excitate the first and to inhibit the second one. Particularly, this means an optimal distribution of the resources amongst the responses.
3. Control of the resources: mobilization (increase of R) or demobilization (R decreases). These reactions can also be called *strain* (activation) and *relaxation*.

Let us consider some of these functions in detail.

First of all we have to distinguish emotions from sensations and perception. The stimulus must be perceived before it can be evaluated as useful or harmful.

In a complicated hierarchical system, sensation is a function of a lower level, evaluation — of a higher level. The objective functions on these levels are different. The higher level takes into account additional conditions which are imposed on the system as a whole. The stimulus perceived can be useful under one condition and harmful under another. Therefore the subject has to differ between the sensation of the stimulus as such, and the «sensation of usefulness» (or harmfulness) of the stimulus. The latter «sensation» is an emotion, a sensation on the higher level of hierarchy. One can say that sensation itself reflects any objective meaning (any object or its feature, stimulus), and emotion reflects a subjective sense — its usefulness or harmfulness for the system as a whole.

Emotions are absent when learning is completed and the stimulus is rigidly connected with the response, i.e., the matrix of conditional probabilities has an ideal form: some probabilities $p(x_i/y_j)$ are equal to one and all others are zero. Mutual information $I(X,Y)$ is maximal and constant and cannot be increased. There is no possibility and no need for the existence of the intermediate evaluation — emotion. Automatized forms of behaviour (instinct, reflex, strong rigid habit, etc.) proceed on the lower level and are usually not accompanied by emotions.

The other important function of emotions is optimization of the behavior. A quantitative estimation of the usefulness of responses, provided by the emotion, make it possible to answer the question: what response — amongst the many possible ones — is the *most* useful, *most* effective? And activation of the most effective response and inhibition of all the others is a form of optimal control based on emotions. Hence, emotion is a mechanism for the comparision of various forms of activity in respect to their effectiveness and distribution of resources in correspondence with their maximal effectiveness. The most effective activity is the first one to receive resources, but the non-effective form of activity does not receive any resources at all.

Intuitively we know what effectiveness and what emotion are. But we seldom connect these things. Let us look at an example to illustrate this connection.

The simplest mechanical example: It is one thing to roll a stone to the top of the mountain when it automatically wants to roll back. And quite another thing to roll it down, when a slight push is enough to move it in the desirable direction. In the second case the effectiveness is much higher and the pleasure more intense. The punishment of Sisyphus was less a useless effort than it was a negative emotion — the inevitable consequence of non-effectiveness.

One can say that various kinds of «amplifiers» (devices which increase the effectiveness of our actions) serve as sources of positive emotion for us. We especially want to pay attention to the numerous technical systems which enlarge the possibilities of man and the effectiveness of his activities. High exactness, speed, and power provided by such technical systems as arms, transport (cars, aircrafts, etc.), computers, and so on — become sources of pleasure for us. On the other hand, any errors and failures in the work of these systems cause a decrease in effectiveness as well as negative emotions.

The last main function of emotions is control of the resources, i.e., decrease or increase of corresponding parameter R. Such emotions as rage, fear, boredom, fatigue, disappointment, etc. are intended just for resource control — mobilization or demobilization of the resources. There are many types of resources and mobilization of one of them can be accompanied by the demobilization of the others. This mobilization can be either general («stress», strain) or local (activation of some response).

Above we have described two types of behaviour — free and compulsory. In free behaviour the subject chooses the answer to the given stimulus in such a way that the objective function L increases. Hence, free behaviour is always accompanied by positive emotions. But what is, then, the source of the negative emotions? The source can be different forms of compulsion — external ones as well as self-compulsion.

It can be any external obstacle which prevents free behaviour of the system. Or the situation of learning, when the teacher forces upon the pupil a new, unhabitual response. Or the necessity to keep a given promise (promise — a form of self-compulsion!). There are many other situations when the subject cannot — or does not want — to change the response. So a situation of self-compulsion is very typical for the human being. Usually the perception of a work of art demands some self-compulsion, especially in the first stage of the perception of any new, strange, exotic work of art. In this case the subject must for some time keep the response fixed regardless of whether it causes positive or negative emotions.

Let us consider how emotion E changes when a new stimulus x_i acts, the system is forced to answer with the response y_j, and probability p_{ij} grows. As we know, these changes pass through three stages:

I. *Rejection.* L decreases with the growth of p_{ij}. We know that in this case the emotion is negative. The subject can simply «switch off» his attention and stop perceiving the stimulus. But if he can compel himself to go on perceiving, then p_{ij} continues to increase and sooner or later the next stage sets in.

II. *Acceptance.* L increases with the growth of p_{ij}. An emotion, we know, becomes positive. In this stage the subject aspires to attract the stimulus, to amplify and to prolong it.

III. *Stagnation.* L decreases. The emotion becomes negative again. The only means to break off this process is to switch to any other, «contrast» combination (x_0, y_0). Then a new cycle of activity begins.

This is a very general regularity of any development. We can find these three stages in many processes of perception on different levels of organization. So the perception of any new idea, scientific concept, or new trend in art usually passes through these three stages: rejection, acceptance, and stagnation. The optimal tactics of behaviour concerning these three stages consists of the following rules:

1. To overcome the first stage as fast as possible (by means of self-compulsion, if necessary).
2. To prolong the second stage for as long as possible.
3. To switch to any «other», «contrast» stimulus at the beginning of the third stage.

Many devices are intended for the solution of these problems. Some of them will be considered in Chapter 5.

2.7 Perception and pattern recognition

Perception is the ability to react to stimuli remote in space (similarly memory is the ability to react to stimuli remote in time). Perception must arise when the system shifts from a homogeneous external media into a heterogeneous one. In homogeneous media there is no necessity to perceive remote stimuli — all features of media (temperature, density, concentration of chemical compounds, etc.) are in direct contact with the organism.

The heterogeneous environment consists of discrete *objects*. An object can be defined as a stable combination of the features that are connected, correlated with each other, and separated from the other combinations (other objects). As a consequence, these objects are separated also from the system, i.e., removed from the system in space. Perception of these objects must be *concrete*, i.e., it must embrace many features of the object *simultaneously*.

If the environment X is described by n binary variables, then $N = 2^n$ states of the environment can exist. But really this number of states is significantly less because of the correlations (internal mutual information $I(X^k, X^l)$) between the stimuli. So, most of the states in the space x remain empty or have low probabilities. These correlations mean that the external environment is organized. We perceive this organization as the existence of the objects, phenomena, regularities, etc.

The existence of these correlations makes the environment more regular, stable, and simple and helps the organism to survive in it. The correlation between the effect and the cause allows the organism to transfer its control from the effect to the cause and provides a more perfect regulation. Correlations between the features of the object permit the organism to judge about the whole on the basis of a part, to recognize and restore the object on the basis of a few features. (This ability plays an important role in the process of creation — e.g., the «recognition» of the solution of the task on the basis of some given features of this solution.) These correlations give rise to the psyche as the ability to reflect on the external environment by the internal structures of the organism.

External organization is reflected within the organism by means of internal organization, correlations between responses, and internal mutual information $I(Y^k, Y^l)$. The simplest example of this correlation is a conditioned reflex. These correlations form the stable combinations of responses which are reflections of the external objects. We call them «notions», «images», «concepts», «models», and so forth. From this moment onwards we can talk about the *object* and the *subject* in a psychological sense.

The formation of these notions is usually called *learning*. The memory is a necessary premise of learning. It stores the result of learning — the joint probabilities $p(x^k, y^k)$ and $p(y^k, y^l)$ which are exactly the model of the situation.

Let x^k be any feature of the external environment, y^k — an internal variable of the system which is connected to x^k by means of any type of physical interaction (light, sound, etc.). Usually the environment has many features $\{x^1, x^2, x^3, \dots, x^m\}$ and the system must also have many internal variables $\{y^1, y^2, y^3, \dots, y^n\}$ to perceive the environment concretely. In general, $m \neq n$.

Let us consider as an example the two rasters X and Y with the pictures x and y on these rasters (see Fig. 1.3 in Chapter 1). Usually we have to consider a set of pictures (vectors, patterns) x which represents various stimuli — external objects, and a set of pictures (vectors, patterns) y — responses on these stimuli.

A particular, simple, but important case is when picture y reproduces (more or less exactly) picture x. We can say that the picture y is an *image* of x. The process of perception in this case looks like a more and more exact reflection, reproduction of x in y, concretization, «enrichment» of the image y by new features of the object x. This means the growth of mutual information between x and y. The maximum of information is achieved when the numbers of degrees of freedom in both rasters are equal and the corresponding cells are in the same states: $x^k = y^k$, i.e., each picture y exactly reproduces the corresponding picture x (Fig. 1.3c). In this case, the entropy $H(X/Y) = 0$ and $I(X,Y) = H(X) = H(Y)$.

But usually many details in object x are superfluous («useless information») and can be omitted in image y. Hence $H(X/Y) = H(X/\overline{Y}) \neq 0$ (see Fig. 1.3b). We say that the image is usually a *generalization* of the object.

A further stage of generalization of the image is *centralization*. Direct correlations between the elements of the image $I(Y^k, Y^l)$ are replaced by correlations $I(Y^k, Y^0)$ with some «central» element y^0. All elements are connected to each other by means of this central element. Usually this is one of the elements of the same level y, that has the maximum of information about other elements and a minimum of resource expense to maintain its connections with these elements. But

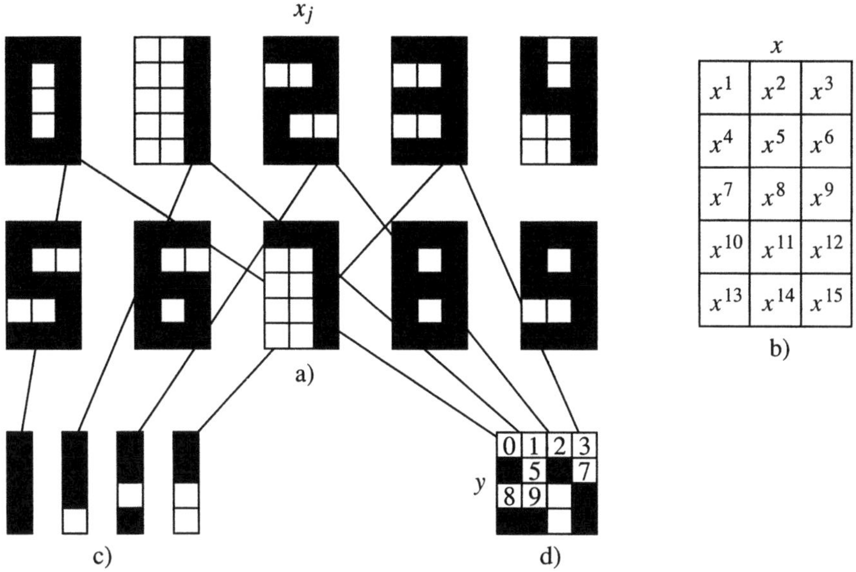

Fig. 2.5

later such elements form a new level z. This is just a higher level of the control hierarchy. It is easy to show that centralization is accompanied by a sharp decrease in the resource expense of the system. Each element y^k has only one connection with the element z of the next level — instead of many contradictory connections with n elements on the same level. Maintenance of the connection demands an average resource expense R. The total number of connections decreases from $n(n-1)/2$ to $n-1$, i.e. $n/2$ times. So the resource expense decreases.

Centralization opens the way for the next important step in evolution — the *symbolical* representation of patterns: the central element does not *reproduce* the pattern, but becomes a symbolical representative of it.

Let us look at an example that demonstrates the difference between the imaginary and the symbolical responses to a stimulus.

In Fig. 2.5a we can see a set of figures playing the role of objects, of stimuli patterns (vectors) x_j. These patterns have been associated with the corresponding responses y_j in the former experience of the subject, during the process of learning. This learning process creates the connections between the stimuli and the responses. These connections don't have to be imaginary, only symbolical: the response (one cell or group of cells in the raster Y) doesn't reproduce the stimulus, but only signifies, symbolizes it. This symbolical connection, the association between stimulus and response is freer and more flexible than the imaginary one and provides more economical and diverse forms of responses.

For example, the patterns $x_0, x_1, x_2, \ldots, x_9$ in Fig. 2.5a we usually associate with symbols «zero», «one», «two»,... ,«nine» or symbols «0», «1», «2»,... ,«9». The set of responses is the set of words or set of figures. But these patterns can be associated also with white cells which occupy corresponding places in raster Y, for instance, the first pattern — with the first cell, the second pattern — with the second cell and so on (Fig. 2.5d). The set of responses is the set of black cells in raster Y. The other possibility is to use binary codes to represent the figures x_j. The set of responses is the set of binary codes «000», «001», «010»,... ,«101» (they are pictured in Fig. 2.5c as combinations of black (0) and white (1) cells), and so forth.

What does the process of perception and pattern recognition look like in this case?

The process looks like the concretization — summing of separate features of the object (components of vector x). Usually the system receives a set of messages. Each of them provides it with information about the state of one cell of the raster X (Fig. 2.5b — white (x_1^k) or black (x_0^k)). This state is one of the features of the stimulus. The system must choose an adequate response on the stimulus on the basis of these features. According to the principle of the information maximum the system chooses as its response the most informative one in the set Y (which has the maximum of mutual information between response y_j and the set of messages received $\{x^1, x^2, x^3, \ldots, x^n\}$).

The initial response is usually very general and «fuzzy». In our case it is a set of figures:

$$Y = \{0, 1, 2, 3, 4, 5, 6, 7, 8, 9\}$$

Each message received gives one of the features x^k of the stimulus pattern and enriches, concretizes this response, and restricts the initial set to the subset of figures containing these features. So, the message «cell x^4 is white» ($x^4 = x_1^4$) restricts the initial set Y to the subset $Y_1^4 = \{1, 2, 3, 7\}$. The next message, for example, «cell x^7 is black» ($x^7 = x_0^7$) causes a further restriction of the subset Y_1^4 to the subset $Y_1^4 \& Y_0^7 = \{2, 3\}$, which includes the figures containing both features indicated. Thus new messages add new features and shorten the initial set until it is reduced to a single figure. For example, the set of messages $\{x_1^4, x_0^7, x_1^{10}\}$ leads to the figure «3». No other figure has such a combination of features. Thus, the process of enrichment and concretization leads to an increase in the «content» of the response (number of features, information) and a decrease in the «volume» of this response (number of figures).

Now we want to describe this process in terms of probabilities. Let us consider a simplest case when features x^k are independent. Each message has to change the old («a priori») probability distribution $(x^1, x^2, \ldots, x^{k-1}/y)$ and supply to it a new multiplier $p(x^k/y)$.

Usually conditional probabilities $p(x_i^k/y)$ are close, but not equal to the ideal values 1 or 0. It means that the set of responses Y_i^k is not exact, but «fuzzy». There is an area of high probability for the «adequate», «normal» responses and

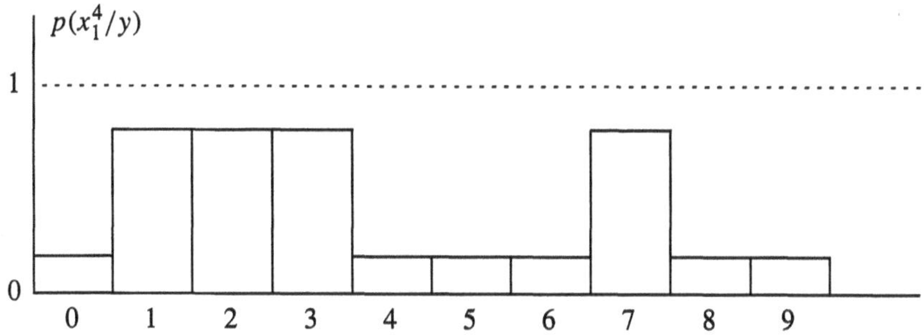

Fig. 2.6

an area of low probability for the unadequate, «deviated» ones. For instance, for the feature x_1^4 the set of high probability is $Y_1^4 = \{1,2,3,7\}$ and the set of low probability — $Y_0^4 = \{0,4,5,6,8,9\}$ (Fig. 2.6).

If we have set of messages with the features $\{x^1, x^2, x^3, \dots, x^n\}$ then the resulting conditional probability distribution can be approximated through the product of partial conditional probabilities:

$$p(x^1, x^2, x^3, \dots, x^n/y) = p(x^1/y)p(x^2/y)\dots p(x^n/y)/Z \qquad (2.23)$$

where $Z = \sum_x p(x^k/y)$ is a normalization factor.

The more n, the more the probabilities $p(x/y)$ approach the ideal values 0 and 1, and the more mutual information between x and y.

Image and symbol are the two main forms that represent the environment in the psyche of a subject.

2.8 Consciousness

Image and symbol, intuition and logic, sense and reason — these oppositions are commonly used when we describe human nature. These opposites have deep evolutionary roots: man belongs to two different worlds simultaneously — the world of Nature and the world of Culture.

The evolution of any living being is directed towards its maximal adaptation to the environment, and to the maximum of information between the conditions of the environment and the responses of the organism. This process obtained a new, important dimension when the human hand, manipulator of objects, appeared. The necessity to increase the variety $H(Y)$ and the exactness $H(Y/X)$ of manipulations led to the appearance of the upright position, liberated forelimbs of man, and transformed them into arms with hands. These new abilities have lead to the creation of *tools* (instruments) and *symbols* (signs) — artificial external objects created to exert control over the environment. The tool controls the physical environment, the symbol — the psychical one. This new type of control interaction becomes profitable due to the properties of tools and symbols that are discussed below.

Tool and symbol became the mediators between man and environment. And this indirect, mediated interaction with the environment becomes the most significant pecularity of man, the one that distinguishes him most clearly from the world of animals. Man receives most information about his environment by means of symbols, and exerts the greater part of his influences on his environment by means of tools.

What are the properties of the symbol that turn it into a more advantageous means of information processing than the image? These properties are:

1. A greater variety of freer connections with the object represented (see Fig 2.5c,d). The symbol can represent a separate feature of the object, which cannot exist without the object; or an abstract notion (concept) which has no obvious, imaginary form at all. It provides an increase in the variety of responses $H(Y)$.

In general, the symbol is a pattern y, different and, as a rule, more simplified than object x, but closely associated with it. Often any typical fragment of the object (as in Fig. 1.3a) or an abstract scheme of the object (as in Fig. 1.3b) becomes the symbol of this object. But this does not have to be so: the symbol can be *arbitrary* and not resemble the corresponding object in any way.

2. The rigid and invariant form of the symbol. This form is often maintained through special social mechanisms (alphabets, dictionaries, grammar, school, normative acts concerning the language, etc.). The work of these external mechanisms is only a manifestation of the process of idealization and is directed by the need to acquire the information maximum. On the other hand, the rigid form of external symbols becomes a great help for the process of idealization in the human mind: in the presence of the external symbol, this process can start from a higher level of information and so demands less resource expense.

The rigidity, «entropylessness» of the symbols permits us to build long chains of reasoning without errors and allows for the existence of such complicated symbolical systems as language, logic, and mathematics.

3. A higher exactness in the representation of quantities. It is difficult to distinguish an image of 99 points from an image of 100 points. But symbols, representing corresponding quantities, are easily distinguished.

Both these properties provide a decrease of the entropy of errors $H(Y/X)$.

4. A more economical use of the resources than in the imaginary representation of the objects.

5. «Reflexivity», the possibility to use a symbol for the representation of the other symbol or group of symbols. The last property permits us to build a hierarchy of symbols, to designate a complicated symbolic structure by a single symbol. The simplest example: it is difficult to distinguish the number *1000000000000000* from *10000000000000000* — we are forced to count zeroes. But thanks to the reflexive

ability of symbols we can designate the number of symbols (zeroes) by another symbol and write: 10^{15} and 10^{16}. These symbols are distinguished easily.

Another example: in Chapter 1 we designated a complicated expression for the average mutual information (1.2a) by the short symbol $I(X, Y)$ and from then on worked with this symbol. This property is widely used in mathematics, physics, and other exact sciences. It permits us to contract the numbers of symbols, to make the reasonings shorter and more obvious, and to economize the resources.

Many properties of symbols are also inherent to the tools: greater exactness and variety of operations; rigidity and, as a consequence, the ability to form large and complicated combinations of tools (machines); economy of resources; and, especially, reflexivity of tools: the tool can be used for making other tools — and so forth.

The important property of tools and symbols, which distinguishes them from the natural organs of animals, is the speed of their development. Tools and symbols as external objects are free from many physical constraints that are inherent to natural organs, such as material, size, form, number, etc. Tools and symbols don't die with man, they can be passed on from one generation or population to another one. Because of this they can accumulate and hand over a collective experience. This collective experience contains a much wider totality of notions, images, habits, and other response patterns than any individual one.

The rapid development of tools and symbols was bound to create a new artificial environment around man. This is the socio-cultural, tool-symbolic environment — civilization. It puts a new problem before man — to adapt himself to this new environment which has different properties than the natural environment. Adaptation to this socio-cultural environment requires new mechanisms and organs of adaptation. The phenomenon of human consciousness is considered to be the general expression of this adaptation.

One of the most important peculiarities of the socio-cultural environment is the speed of its development, faster than the evolution of the natural environment. This circumstance must require the same rate of development from the «human» compartments of the brain. In principle, Nature could in this case use the well-known way of building up a «super-cortex» over the cortex (like the cortex itself had been built up over the undercortex). This has partly been accomplished in the development of the frontal and temple departments of the cortex.

But on the whole, the speed of the morphological evolution was insufficient and the problem was mainly resolved in another way. One of the two hemispheres of the human brain (which duplicate each other in animals) was built up — in a functional sense — to stand over the other one and became an organ for adaptation to the socio-cultural environment, an «organ of consciousness». So the powerful development of the interhemispherical nervous connections, so typical for man, is apparently a morphological manifestation of a new, «vertical» type of hemispheric relations. We have to add that the former, «horizontal» type of relations remained the cause of additional complications in the picture of hemisphere interrelations.

The idea that the left hemisphere is the «organ of consciousness», the organ of adaptation to the tool-symbolic environment allows us to explain the «binary» nature of a human being and the principal differences in the functions of both hemispheres. These differences can be enumerated as follows:

a. The symbol is the instrument of the left hemisphere, and the image is the instrument of the right one. Language, numbers, and abstract concepts are the instruments of the left hemisphere.

b. The analytical, abstract, and formal character of the operations performed by the left hemisphere in contrast to the concrete, synthetical, and imaginal character of the operations performed by the right one. Formal, symbolic methods of logic and mathematics are functions of the left hemisphere.

c. The right hand (controlled by the left hemisphere) has the leading role in operations with tools and symbols (working, writing, etc.).

d. The performance of the function of reflexion, of self-consciousness (self-perception of the psychical processes, self-evaluation, self-control, self-compulsion, self-punishment, etc.) is executed by the left hemisphere.

The last property is connected with the reflexive, «self-convertible» nature of symbols and tools: a tool can be used to produce other tools and a symbol — to signify other symbols. This property permits the increase of the effectiveness and rate of evolution of the socio-cultural environment compared to that of the natural environment.

Usually reflexion is understood as «knowledge about knowledge», as the ability of the subject to realize the means and foundations of his activity, i.e., as a property of the human consciousness. But it is only the particular case of reflexion as a manifestation of the more general property of the system in evolution, which was defined in the beginning of this chapter. According to this definition, reflexion is not only self-perception and self-evaluation, but also self-control, self-compulsion, and self-punishment.

Let us look at some examples of reflexion and its influence on human activity.

1. The first example concerns the human emotions. Above we have determined emotions to be the means of optimal control of behaviour, as a new level of reflexion over the sensations and responses. But in the next stage of reflexion, emotions are transformed from the means into the end. For an animal, emotions — pain and pleasure — are not more than means which direct behaviour to its aim. The aim is to adapt, to fit the animal into the environment. Human emotion becomes in many ways a self-purpose, giving rise to such a human phenomenon as «hedonism» — the striving for pleasure for its own sake.

Human love and human hate are also transformed into a self-purpose, an end in itself. The tiger kills for food, he doesn't hate his victim. Only man kills for pleasure, transforms murder into the art, into an end in itself. In the same way, sex for an animal is only the means to procreate, continue the species. Only man transforms the sex into an end in itself, into the refined art of love. Food for man is not merely a means of getting rid of hunger, but also a means of obtaining various delights.

The art of cookery, love and hate as art, science as art, and in the end the «art for art's sake» — those are the fruits of reflexion.

2. Reflexion also transforms the human cognition, thus giving rise to forms of cognition like science. Science arises when the truth becomes the end in itself, independently of its practical significance. Ancient geometry and ancient astronomy were a means for solving practical problems. But later the study of the celestial bodies and the properties of the figures became an aim in itself.

The higher type of scientific knowledge — scientific theory — is also the product of reflexion. As we mentioned in the «Introduction», a «genuine» scientific theory must have two levels:

a. Empirical knowledge, when facts are connected (and thanks to this, explained) by means of laws. Even when these laws are in a mathematical form, they are not a real theory yet. In principle, this form of knowledge is accessible even for animals — in the form of reflexes, associations, etc.

b. Theoretical knowledge, when reflexion transforms the laws itself from the means of explanation to the object of explanation. This explanation is reached by logically connecting these laws with other laws or with any general principles.

In Fig. 2.7 we see a part of the second (logical) level of mechanics. (This picture is based on the textbook of mechanics by Landau and Lifshitz, 1958.) Each new particular sentence is deduced from the previous (more general) one and some supplemental condition (these conditions are placed on the left side of the picture). The main logical construction that forms the structure of this level is:

$$IF\ X^i\ AND\ X^j\ THEN\ Y.$$

Here X^i and X^j are premises, Y — conclusion. In the next step the conclusion becomes a new premise, $Y = X$ — and so on. As is obvious from the picture, the initial, central, and most general premise is, of course, the variational principle. In this case — the principle of the least action: $\delta S = 0$.

3. The last example: according to Max Weber (1965), the «spirit of capitalism» arises when «profit» is transformed from the means of satisfaction of the needs into an end in itself: «Profit becomes an end in itself to such a degree that it is changed into something transcendent or even simply irrational in respect to the «happiness» or «usefulness» of man. Now, it is not acquisition that serves man as a means of needs satisfaction, but the whole existence of man is geared towards acquisition, which becomes the aim of life.»

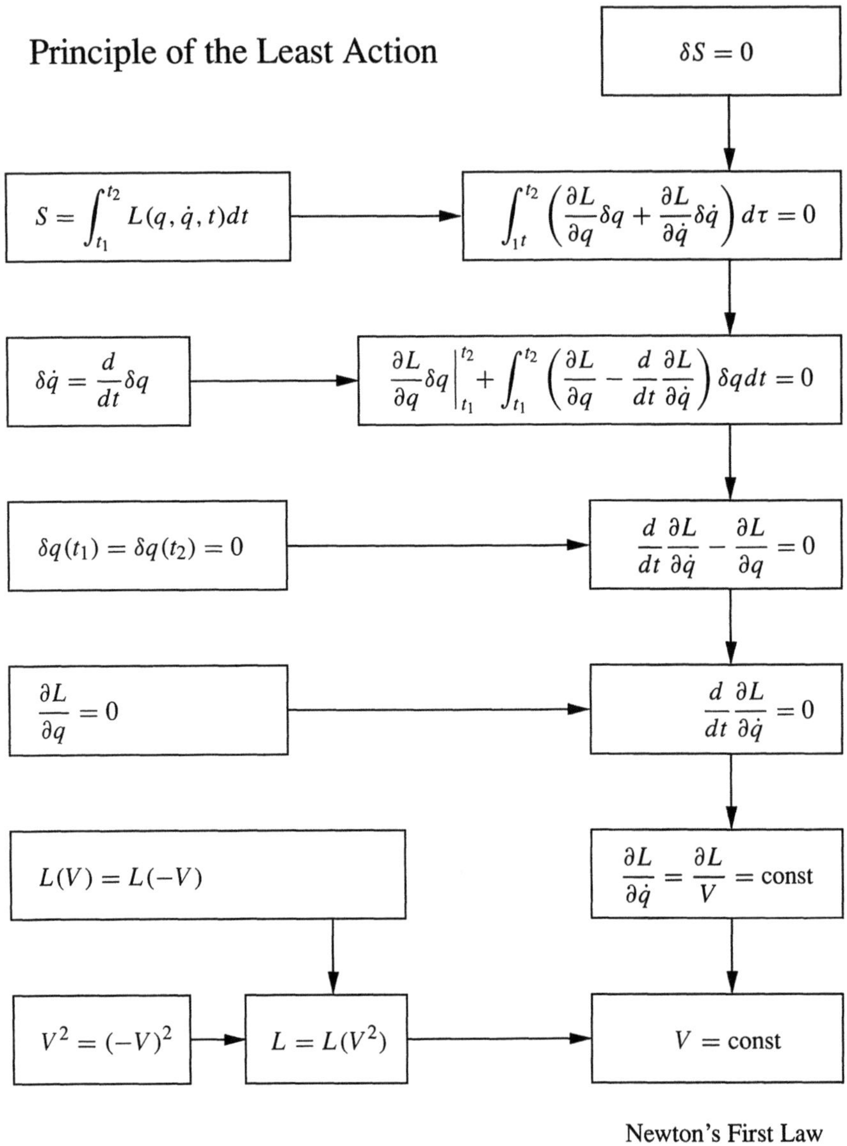
Principle of the Least Action
$\delta S = 0$
$S = \int_{t_1}^{t_2} L(q, \dot{q}, t)dt$
$\int_{1t}^{t_2} \left(\frac{\partial L}{\partial q}\delta q + \frac{\partial L}{\partial \dot{q}}\delta \dot{q} \right) d\tau = 0$
$\delta \dot{q} = \frac{d}{dt}\delta q$
$\frac{\partial L}{\partial q}\delta q \Big|_{t_1}^{t_2} + \int_{t_1}^{t_2} \left(\frac{\partial L}{\partial q} - \frac{d}{dt}\frac{\partial L}{\partial \dot{q}} \right) \delta q dt = 0$
$\delta q(t_1) = \delta q(t_2) = 0$
$\frac{d}{dt}\frac{\partial L}{\partial \dot{q}} - \frac{\partial L}{\partial q} = 0$
$\frac{\partial L}{\partial q} = 0$
$\frac{d}{dt}\frac{\partial L}{\partial \dot{q}} = 0$
$L(V) = L(-V)$
$\frac{\partial L}{\partial \dot{q}} = \frac{\partial L}{V} = \text{const}$
$V^2 = (-V)^2$
$L = L(V^2)$
$V = \text{const}$
Newton's First Law

Fig. 2.7

2.9 Thinking

Image and symbol are the two main forms that represent the environment in the psyche of man. The symbolical level of representation is a new level with new elements — symbols, and new connections between the elements. These connections are usually named «syntactical» or «grammatical» ones. One of the typical and important examples of such connections are logical operations. They become a basis of thinking.

Symbolical (particularly logical and mathematical) operations emerge as a new, more effective means for the solution of human problems. A property of the symbol such as rigidity provides a compulsory idealization of human notions. An arbitrary connection between the symbol and the notion permits us to use abstract features and notions in thinking. And the last, but not least, important property of the symbol is the ability to form new types of connections between the symbols — grammatical (or syntactical) ones. These properties of the symbol became the basis of such symbolical systems as language, logic, and mathematics, which became a powerful means of self-control, self-compulsion, and solving of problems. We shall look at language in detail in Chapter 3. Here we want to look at a particular case of the logical (and partially mathematical) language methods.

Logic (side by side with perception) is, in essence, the means of the formation of notions. But they can be essentially new notions that have not been perceived before. Or, even, that cannot be perceived or imagined at all!

But logic arises on the basis of operations of the information performance that are inherent to perception. Logical operations are the extreme, ideal case of the informational operations. This idealization, as we know, leads to an increase in the mutual information. Therefore, it is a further natural development of the informational operations and is subordinate to the principle of the information maximum.

We know that the human brain is a thermodynamical system and all the spontaneous processes in such a system are irreversible and must be accompanied by the growth of entropy. Then what is the reason of this reversibility, «entropylessness» of the logical operations, performed by the brain? One of the reasons is the fact that man uses the tools and symbols, especially «external» «hard» copies of these symbols (e.g., letters printed on paper, etc.). But these external means play only an auxiliary role. The main reason is the aspiration of the human mind itself to the ideal forms, an «idealization» which is a consequence of the principle of the information maximum. External means only help the human mind to maintain an ideal form of informational operations. They are not ideal in themselves, but they provide a higher initial «level of idealization» than the natural mechanisms of the human brain and make it easier for man to achieve the ideal state.

For instance, a real point on this page is not an «ideal point» of geometry, it is only a symbol of this ideal point; but it helps our mind to approximate this ideal.

In the process of idealization, probabilities p_{ij} which lie over the threshold increase, and probabilities which lie under the threshold decrease. Thus large probabilities become larger, small probabilities become smaller and approximate zero, and the matrix of probabilities as a whole aspires to the ideal form. Such idealized probabilities can be treated as logical functions — predicates, and their combinations can be described by means of logical operations.

The predicate is a logical function which has a value «true» or «false», depending on what value of y is taken into account. So the statement «white cell x^4 is a feature of y» (in short, $P(x_1^4/y)$) is true for the figures $y = 1,2,3,7$ and false for the other figures (Fig. 2.5). The set $\{1,2,3,7\}$ may be called also the «solution set» (or «set of solution») for the statement $P(x_1^4/y)$.

Note, in this expression $P(x_1^4/y)$ feature x_1^4 is a fixed constant and y (in logic this is often called the «subject») is a variable which can change in value.

Let us signify the «true» significance of the predicate as *1* and the «false» significance as *0*. Then one can see that the predicate $P(x^k/y)$ is the limit, the ideal case of the probability distribution $p(x^k/y)$, when the «weight» (coefficient of amplification) n of the stimulus x^k approaches infinity:

$$P(x^k/y) = \lim_{n\to\infty} p(x^k/y)^n/Z \tag{2.24}$$

where $Z = \sum_k p(x^k/y)^n$.

This correlation between P and p permits the explanation of one interesting fact. From the point of view of the «pure» logic, the simple repetition of a statement is useless: this repetition brings no new information and does not make the statement more true (or more false). At the same time, we know that in real life simple repetition of a statement is often useful, it brings new information and makes our notion more (or less) probable. This repetition is often necessary to affirm this statement as true or false in our mind. This difference becomes intelligible if we assume that the predicate is a limit of the conditional probability when the experience on which the probability is based (weight, number of observations, repetitions n) appoximates infinity. In other words, pure logic deals with statements, sentences, and predicates which are based as if on infinite experience. In this case, any finite number of repetitions will not increase this infinity and, consequently, has no sense. Of course, in real life, infinite experience is an impossibility.

Now let us consider a logical process of the formation of a notion. It can be treated as the idealization of the process of notion formation in perception.

A main operation, which is common for perception and logic, is concretization. Usually the formation of a notion in logic looks like a combination of separate conditions and features of the object, and a consequent supplying of these features to the abstract initial notion, i.e., concretization.

From the point of view of logic, the concretization of the response is described as a conjunction of predicates and as an intersection of the corresponding solution sets. So the statement $P(x_1^{13}/y)$ («x_1^{13} is a feature of y») has the solution set

$Y_1^{13} = \{1,4,7\}$, and the statement $P(x_0^4/y)$ («x_0^4 is a feature of y») has the solution set $Y_0^4 = \{0,4,5,6,8,9\}$. The conjunction of these statements $P(x_1^{13}/y)$ *AND* $P(x_0^4/y)$ («x_1^{13} and x_0^4 are features of y») has the solution set which is the intersection of these two sets and contains the single figure — «4»:

$$Y_1^{13}\&Y_0^4 = \{1,4,7\}\&\{0,4,5,6,8,9\} = \{4\}.$$

If we signify the «true» significance of the predicate as 1 and the «false» significance as 0, then we can describe the result of the concretization, i.e., the complicated («many-place») predicate $P(x^1, x^2, \ldots, x^n/y)$ as the product of the simple («one-place») predicates:

$$P(x^1, x^2, \ldots, x^n/y) = P(x^1/y)P(x^2/y)\ldots P(x^n/y). \tag{2.25}$$

The solution set of a complicated predicate is the mere conjunction of the solution sets of the simple predicates. Thus the multiplication of the conditional probabilities (2.23) is replaced by the logical conjunction of predicates.

The second important operation of the notion formation is idealization. «True» and «false» are idealizations of the notions «of high probability» and «of low probability» correspondingly.

We see the logical description of the response formation is an idealization of reality. It assumes the absence of noise and errors in the informational «channel» connecting x to y.

The process of notion formation can be described now by means of logical operations, such as *AND, OR, NOT, IF... THEN*. So we can write for patterns in Fig. 2.5:

IF x_0^1 *AND NOT* x_0^4 *THEN* 2 *OR* 3 *OR* 7;

IF x_1^{13} *AND NOT* x_1^4 *THEN* 4 — and so on.

Here *NOT* x_1^k is equivalent of x_0^k.

A more complicated example of logical connections is Fig. 2.7 where we present a successive deduction of the laws of mechanics from the initial postulates and supplemental conditions.

2.10 Creation

What is creation? What is a creative task?

«To create» is first of all «to make something new». In the language of probabilities «new» means «of low probability» and «old» means «of high probability».

Furthermore, «something new» is usually a new combination of old elements. So, a new combination of «this» stimulus x_i and «this» response y_j means that the probability $p(x_i, y_j)$ is small, but the stimulus itself and the response itself do not have to be new, i.e., probabilities $p(x_i)$ and $p(y_j)$ can be large enough! Hence, $p(x_i, y_j) < p(x_i)p(y_j)$ and information

$$I(x_i, y_j) = \log \frac{p(x_i, y_j)}{p(x_i)p(y_j)} < 0 \tag{2.26}$$

is negative. We know that in this case, «wrong habits» exist — high probabilities to answer with an «other» response to «this» stimulus $p(x_i, y_0)$ and with «this» response to an «other» stimulus $p(x_0, y_j)$.

Hence, «new» is not always «informative» — in the first stage, «new» usually has a negative informativeness and must overcome some threshold to become «informative». A new combination of stimulus and response must be fixed, maintained, and repeated up to the end of the first stage to overcome the threshold. This fixation can be maintained by means of any external cumpulsory factors (for instance, by the teacher in the process of learning).

But creation emerges when the *self-compulsion* of the subject takes over this role. The subject continues to fix the «unprofitable» stimulus x_i in spite of its informativeness remaining negative. In other words, creation — in the exact sense, as an active process — emerges together with human consciousness and with the human ability of self-compulsion.

As a consequence, human creativity looks like the solving of any «creative task»: a «new» arises when a certain task is completed and certain contradictions have been removed. The conditions of the task can be seen as the stimulus, the solving of the task — as the response. But many tasks can be solved through existing, regular, standard methods and cannot be considered to be creative ones. Thus a definition of the creative task becomes necessary.

The most general definition of a genuine creative task is a task that demands us to overcome any «wrong habits» — prejudices, predispositions, old notions, etc. As we know, these habits form a threshold which prevents us from forming an adequate response y_j (solution of the task) to the given stimulus x_i (conditions of the task). Hence, creation is seen not only as the making of something new, but also as the *overcoming of something old*. The solving of the creative task can be seen as the change of inadequate habitual response y_0 and formation of an adequate response y_j on given conditions.

For example, one of the prejudices that may arise on the basis of the figures 0...9 (see Fig. 2.5) is that the cell x^3 is always black. This conviction becomes false if we pass from the set of figures y_0 to the set of letters y_j. We see that the letter «L» has a white cell x^3. Such habits and prejudices exist usually because the experience of each subject is always restricted.

The necessity to overcome the threshold formed by old habits is a trait of the «genuine» creative task and creative process. It is precisely the overcoming of this threshold that has been labelled in psychology as «insight», «creative leap», «break», «illumination», and so on. And this step over the threshold, moving from

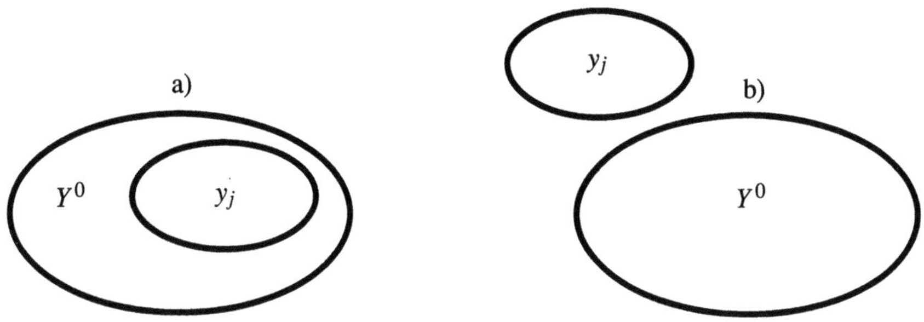

Fig. 2.8

the first stage to the second one, causes the positive emotions in the *second* stage of the creative process.

As we know, a compulsion is necessary to overcome the threshold. Therefore, compulsion and negative emotions in the *first* stage are also traits of the «genuine» creative process.

In human creation, something «new» usually emerges as a new notion: it can be a new concept («field» by Maxwell, «natural selection» by Darwin), a new image («Hamlet» by Shakespeare, «Faust» by Goethe), a new mathematical formula («$S = \log W$» by Boltzmann, «$E = mc^2$» by Einstein), a sketch of a new machine, a new chess combination, a new architectural design and so on. Hence, regularities of the formation of a notion (see Figs. 2.5 and 2.9) are the regularities of the creative process.

Usually a new notion has to be formed with the old one in the background. This old, habitual notion can be very general and «fuzzy», so that the subject will not realize it exists. But this subconscious background notion always exists and influences the formation of the new one. The process of emergence of the notion essentially depends on the relations existing between the new notion and the old one. Here we must make a distinction between two sharply different situations:

1. A new notion y_j coincides (fully or partially) with the old one (Fig. 2.8a). For instance: the old notion is the set $Y^0 = \{0, 1, 2, 3, 4, 5, 6, 7, 8, 9\}$ and the new one is the figure «3» which coincides with a part of the old notion.

2. A new notion y_j falls out of the boundaries of the old one (Fig. 2.8b). For instance the old notion is Y^0 and the new one is the letter «L». This is a genuinely new notion.

Now we can define a creative task in these terms: a task is creative for a certain subject if the solution of the task in given conditions is a new notion and falls out of the boundaries of the old, habitual notion of the subject. In other words, the old notion is wrong. Since different subjects have different habitual notions, this definition is relative: the task may be creative for one subject and trivial for another one if they have different habitual notions. As a result of the solution,

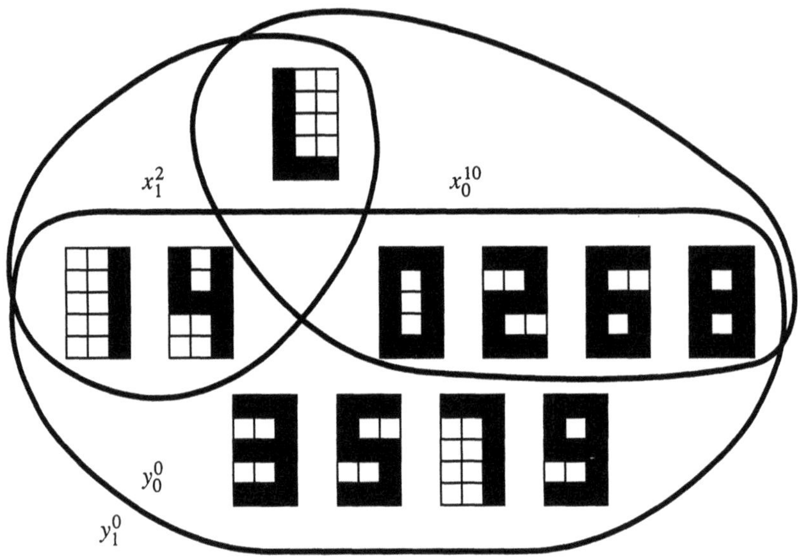

Fig. 2.9

the set of notions of the subject is expanded so that the new notion becomes a habitual one and the task ceases to be creative.

We can also define a creative task in terms of sets: the task is creative if the features of the solution x^k, which are provided by the conditions of the task, are contradictory, incompatible, i.e., the intersection of the corresponding solution sets Y^k is empty:

$$Y^0 \& Y^1 \& Y^2 \& \ldots \& Y^n = \emptyset. \tag{2.27}$$

Note that the old, habitual set Y^0, without fail, participates in the conjunction of the sets. This is a very important fact because this habit restricts the circle of the notions of the subject, and makes the task creative. The intersection of the solution sets is empty *only within the boundaries of this habitual set.* If the boundaries of this set are expanded, the situation can be changed and the contradiction can be resolved.

Example: in Fig. 2.5 the simultaneous presence of the features x_1^2 («cell x^2 is white», solution set $Y_1^2 = \{1,4\}$) and x_0^{10} («cell x^{10} is black», solution set $Y_0^{10} = \{0,2,6,8,9\}$) seems to be impossible. But this contradiction disappears if we move outside the habitual set {figures} and consider a wider set, for example — {figures + letters}. In this expanded set, the solution sets can intersect (Fig. 2.9) and one can find the object that has both features indicated. It may be, for example, the letter «L».

Thus we can give another definition of a «creative task»: the task is creative if the old set of notions is restricted and doesn't contain the new notion y_j with the features x_i indicated by the conditions of the task. In other words, the creative task is the task which demands *expansion* of the old response set in order to embrace the solution.

We can also define a creative task in terms of predicates: if the conjunction of predicates describing the features of the solution (given by the conditions of the task) is false

$$P(x^0/y)P(x^1/y)\dots P(x^n/y) = 0 \tag{2.28}$$

then the task is creative. Note that in the series of predicates we must include a fictitious predicate $P(x^0/y)$ which is not found among the given conditions of the task, but describes the habitual, old notions of the subject and is unconsciously added to these conditions by the subject. In the end we can give the most general definition of the creative task in terms of probabilities: if the conditional probability corresponding to the features of the solution is low (i.e., the solution is a new notion!) then the task is creative:

$$p(x/y) = p(x^0/y)p(x^1/y)\dots p(x^n/y) < \bar{p}. \tag{2.29}$$

Here $\bar{p}$ is a threshold probability. It can be connected with the expression for the threshold (2.20) as follows:

$$\bar{p} = \bar{p}(x/y) = \bar{p}(x,y)/p(y) = \bar{p}_{ij}/p_j = \bar{p}_{ij}/(\bar{p}_{ij} + p_{0j}) \tag{2.30}$$

where $\bar{p}_{ij}$ is derived from (2.20).

Note, in analogy to (2.27) and (2.28) we have to include in the series of probabilities an initial, «a priori» probability $p(x^0/y)$ which characterizes the set of old notions of the subject. Whenever these notions do not contain a solution, the probability of solution will be low (i.e., lies under the threshold). This makes the solution a new notion, and the task — a creative task.

Let us look at some examples of real creative problem solving.

1. The famous «Four points» problem (Ya.A. Ponomarev, 1976). It is necessary to connect four points by means of three rectilinear segments composing a closed figure (Fig. 2.10). The fixed features of the solution, set by the conditions of the task (stimuli x^k), can be described as follows:

x^1 — to connect *four* points;

x^2 — by means of *three* rectilinear segments;

x^3 — composing a *closed* figure.

Each feature comes up with a corresponding solution set which is shown in Fig. 2.10. But, as a rule, the intersection of these sets is empty, i.e., does not contain any figure that satisfies all these conditions. In other words, these features look contradictory, incompatible to each other. This happens because, as a rule, the subject has a restricted view due to the strange unconscious prejudice that he can only use as responses the *short* segments (x_0^0) — with a length equal to the

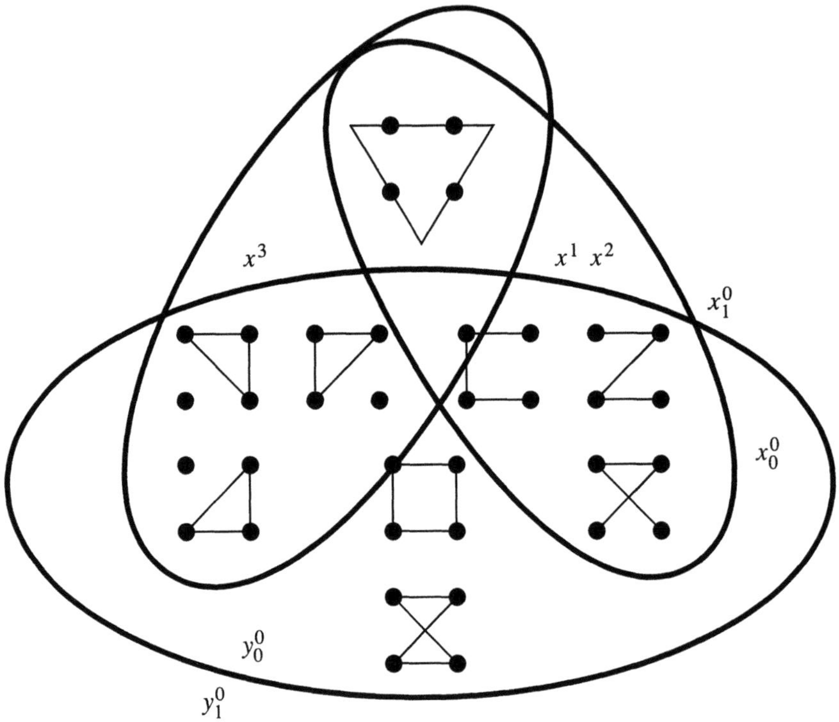

Fig. 2.10

distance between the points. As a consequence, the set of the responses Y_0^0 used by the subject is restricted by the short-segment figures and does not contain the solution.

The solution is found immediately, as soon as the subject realizes his prejudice, moves out of the old set of notions, and begins to use *longer* segments (x_1^0).

2. Let x be a fixed parameter (stimulus) of $x = 2$. Whether or not it is possible for us to solve the equation $y^2 = x$ depends on our notion of the permissible responses (numbers) y and their corresponding sets of solutions. If we restrict ourselves to the set of *natural* numbers (Y_0^0), this equation has no solution. We must overcome some of our predudices about the features of the numbers, and accept the existence of *fractional* and *irrational* numbers (Y_1^0), i.e., expand our old set of responses. In this case we can find the solution which is the irrational number $y = 1.41\ldots$

But the slighly changed equation $y^2 = -x$ (where $x = 2$) has no solution even in this extended set of numbers. And we need a further expansion of this set to the set of complex numbers.

The whole history of mathematics is the history of solving creative problems; and the whole history of mathematics is the history of the progressive expansion of our notions about mathematical objects (numbers, functions, operations, etc.) which are possible solutions to these problems.

3. At the end of the 19th century, two contradictory notions were formed in physics on the basis of a great amount of facts:

x^1 — All the inertial reference frames are *equivalent.*

x^2 — The speed of light is *the same* within all inertial reference frames.

Albert Einstein has shown that a contradiction can be resolved if we throw out our habitual notion about *absolute* time (Y_0^0) and consider time as a *relative* quantity (Y_1^0): in different reference frames the time flows in different manners.

4. Each artist must solve a creative task when he forms a new image. This image must do away with old notions that have formed in the consciousness of the spectator (reader, listener, etc.) by his former experience. This experience includes, particularly, previous images and notions created by the artist himself: each work of art is a totality of consecutive and/or simultaneous images, notions, and patterns which form a background for the new image and influence its perception. Many of the aesthetic devices and methods see it as their only purpose to overcome the threshold created by this background and to replace an old notion with a new one.

2.11 Intuition and logic

As soon as we have a definition of the creative task, we can find devices and methods to solve it. These devices are devices of the notion formation and threshold overcoming.

As we have found above, two main processes of the notion formation exist:

1. *Concretization* — enrichment of the response pattern by new features and the formation of the joint conditional probability $p(x^1, x^2, \dots, x^n/y)$. Usually this is a compulsory process which supplies new conditions $R(x^k, y)$ to the objective function and adds new items $p(x^k/y)$ to the old probability distribution. It allows us to by-pass the first stage and overcome the threshold.

2. *Idealization* — the spontaneous process (the second stage) when the conditional probabilities $p(x/y)$ aspire to ideal values 1 or 0.

Usually both these processes are present in the perception of the features, and alternate with each other in various order. The two extreme cases of this order are the most interesting:

1. All the features are perceived simultaneously so that we have at first concretization and then the overcoming of the threshold and the idealization:

$$p'(x/y) = \mathrm{I}\prod_k p(x^k/y)/Z. \tag{2.31}$$

2. The features overcome the threshold and are perceived consequently, so that the idealization of the partial conditional probability distribution has finished before the other feature is perceived:

$$p''(x/y) = \prod_k \mathrm{I}\, p(x^k/y)/Z. \tag{2.32}$$

In the second case we deal with a *logical* way of response forming. The first one can be identified with *intuition*. Above we have connected these two ways with the left and right hemispheres of the human brain. Both ways permit an increase in the weight n and the probability of a new notion, and to overcome the threshold and find the solution, but by different means.

The conditions of the threshold overcoming (2.20)

$$p_{ij} - \frac{p_{i0}p_{0j}}{p_{00} - 1 + \exp[-I - \beta(r_{ij} - R)]} = 0 \tag{2.20}$$

are precisely the conditions that stimulate intuition and can be considered to be «psychological» devices of creation. Let us remember some of these conditions in relation to the creative process:

a. «Sensibilization» (increase of the probability p_{ij}). This is a very important device, but it seems so simple and obvious that usually the subject pays no attention to it. It is very important to fix as rigidly as possible in the consciousness the conditions of the task x_i. Observation of the work of inventors shows that often they forget to fulfil it.

b. Absence of «wrong habits» — predispositions, prejudices, etc. (lowering probabilities p_{i0}, p_{0j}). The conditions of this task x_i must not be connected with the other solutions y_0 and vice versa. It is one of the reasons why a professional with rigid habits cannot solve a problem which is easily solved by a novice.

c. Switching on the other, contrast activity (increase the probability p_{00}). It often occurs that if the subject is distracted by another activity and then returns to the task, he often finds the task is solved. Rest is a particular case of this switching.

d. Removal of factors, which distract the attention (lowering the average informativeness of the background I). Particularly, the «core» of the task must be cleaned and presented in its most compact and obvious form. This compactness is a good means to stimulate intuition.

e. «Inspiration» — mobilization and concentration of the resources (lowering of the deficit of the resources β).

f. Facilitating the task (decrease of the resource expense r_{ij}). This means removal of all superfluous, unnecessary factors that demand resources, but do not help the solution.

g. Concreteness. Intuition uses a simultaneous, «parallel» account of the conditions given x_i, and the success of intuition in high degree depends on the completeness of this scope. The more features of the solution are embraced, the «richer», concreter the notion about this solution, the lower the threshold, and the easier the «informational contact» with the solution arises. The large weight of the solution is a result of the large amount of different, but simultaneously perceived, features. This simultaneous combination of features forms a concrete *image*. Thus the image, imagination, and imaginary thinking is a main means of intuition in the solving of creative tasks.

But this means has shortcomings which forced man to develop a new, powerful means for problem solving. The basis of this means became the left hemisphere of the human brain with such tools of the information processing as concept, symbol, and logical and mathematical operations with symbols. We shall unite all these means under the general name of «logic».

In contrast to intuition, the means of logic are *idealization* and *abstraction*. As we know, idealization is an equivivalent of infinitely large weight. Thus, concreteness in logic is not necessary to overcome the threshold, so logical sentences may be *abstract*. Hence, abstract concepts and predicates are the tools of logic. An important property of these concepts is the possibility to designate them by means of symbols; and to use a new type of connections between the symbols — syntactical ones — to solve the task.

The abstractness of logic is not a shortcoming, but an advantage, which provides a more effective solution to the creative task. It is the new step in the development of human thinking. Let us consider this topic in detail.

We have found that a concrete image can help the subject to overcome the threshold and solve a creative task. But, paradoxically though it may seen, the image can also prevent the subject from finding the solution. Indeed the image increases the concreteness and the weight of the notion. And it is good if the notion is right. But if it is wrong? We know it is typical for the creative problem that the subject is entrapped by the wrong habitual notion. And it is necessary to *decrease* the concreteness and the weight of this notion to solve the problem.

The image is a concrete notion. That distinguishes it from the concept — the abstract notion. So, in our example (Fig. 2.5) figure «3» is the concrete image. And set $Y_1^4 = \{1, 2, 3, 7\}$ (including figure «3») — the abstract concept. But the danger is that the image every now and then is used as a «typical representative» of the class (set) of objects, i.e. substitutes the abstract concept corresponding to this class. The image and the concept correlate with this class in different ways: the concept includes all objects (elements) of the class with various values of their features. The image includes only one object (or one type of objects) with one value of each feature.

So, the concept of a triangle includes acute, right, and obtuse triangles; the image of triangle can correspond to only one of thess types, for example, to be right triangle. The other values of the feature (and the corresponding objects) are excluded, which leads to the actual contraction of the set of objects considered. On the other hand, we are usually inclined to use the concrete image of the triangle as the «triangle in general», as a representative of the whole class of triangles.

As a consequence, we unconsciously attribute particular features of the concrete triangle to all triangles. This can become a source of errors.

As far as the triangle is concerned such error can today be made only by a young student. But let us remember the trouble that was caused to mathematicians by the image of the function as some sufficiently smooth curve, and how long this notion prevented us from solving the problems in the calculus of variations and in other areas of mathematics. In the end, mathematicians were forced to reject this notion. Today a function looks like a very abstract concept which often can be neither pictured nor imagined. For example, the modern function may have infinite numbers of ruptures in the finite segment, «splashes» with infinitely large height and infinitely small width, etc.

Thus «image thinking» becomes dangerous when it relates not to the solution but to the conditions of the task, which determine a set of the variants considered. Solving the creative problem demands the extension of this set, but image thinking leads to the contraction of it, risking the loss of the solution.

There exists a joke illustrating the dangers of image thinking. Once the famous physicist P. Kapitza proposed to other physicists, L. Landau and L. Infeld, the following task: «One ties the frying pan to the dog's tail. When the dog runs, the frying pan bangs on the road. The question is: with what velocity must the dog run, to make us not hear the bang of the frying pan?» Two famous physicists wasted their efforts but could not solve the task. At last Kapitza took pity and prompted the solution: velocity must be *zero*.

In this example we see how the intentional imageness of the task conditions leads to the contraction of the solution set and to the loss of the solution: the *image* of the *running* dog excludes the zero value of velocity. Note that the *concept* of velocity doesn't know this restriction, it allows not only zero but even negative values.

We see a similar effect in the problem «Four points»: bright black points insidiously impose on the subject instead of the *concept* of a segment with the arbitrary length — a quite defined *image* of a segment with a length equal to the distance between the points.

Thus the wrong notion, which has a concrete image form, can become the main obstacle in solving the problem. This conclusion is confirmed by many facts which demonstrate that in the technique of problem solving, the tendency to reject the image forms for the symbolic ones is observed. Such evolution — from the obvious notions to the abstract concepts, from the images to the symbols — is typical for the many natural and artificial languages. We can remember a transition from pictographical to ideographical writing. Or the evolution of geographical maps: mountains, towns, forests, and animals living in them were formerly pictured in maps. Now we see only abstract conditional marks which are uncomprehensible without an explanation. Such evolution has taken place comparatively recently in the language of radioschemes: obvious pictures of radiolamps, coils, condensers, etc. were replaced by conditional symbols which bear litle resemblance to these objects.

But such development doesn't reject human intuition — this intuition only switches to a higher, symbolical level of the hierarchy and remains the paramount component of human intellect. Often logic is used only as a means to transform the task to such simple and obvious form that the solution can be found directly, on the basis of intuition.

A new control level — consciousness — provides man with an opportunity for control, for compulsory alteration of the old set of notions. Man can force himself to take into account a new notion that has a low probability and is not informative. The means of self-compulsion are symbols, symbolical operations, and symbolical systems, such as language, logic and mathematics, that prescribe rules for these operations. For instance, a logical operation like the *negation* moves the subject from the old set of notions Y_0 to its supplement Y_1. In the case of a creative task this can immediately lead the subject to the solution. The history of mathematics, physics, and other science and technology shows us many examples when the rejection of one of the habitual, old features of a solution makes it possible to solve the problem. For many years, Hamilton searched unsuccessfully for any analogy of complex numbers in three-dimensional space. The solution was immediately found when he realized and rejected such a habitual and almost unconscious feature of numbers as *commutability*.

Symbolic methods of logic and mathematics are powerful means of self-compulsion and self-control. They discipline the capricious and erroneous human mind and prevent it from deviating from the true path under the influence of such disturbance factors as prejudices, emotions, tiredness, and so on. There is a new type of link between the symbols — grammatical (syntactical) ones. They are quite different from the correlations between notions, as well as from the semantical associations between the symbols and notions (see Chapter 3). Thus, the transition on the symbolical level releases man from the habitual correlations between the notions. These correlations often predetermine the motions of human imagination and prevent it from finding an unusual solution for the creative task. Grammatical links can lead man in a direction he could not have chosen on the basis of his imagination only.

For instance, many times in the history of science the scientist, extracting a square root, received — besides the positive meaning — also an unbidden negative one. And involuntarily began to think about its physical sense. So, the expression for energy in the theory of relativity looks like:

$$E = c\sqrt{m^2c^2 + p^2}. \tag{2.33}$$

The formula shows clearly that besides the positive meaning of energy, a negative one is also possible. But physicists, having their habitual notions about energy, for a long time rejected this conclusion as physically senseless. It was only

P. Dirac who believed mathematics and accepted this result. We know it has led to the discovery of the positron and the giant world of anti-particles.

Another example: W. Heisenberg was very troubled when he discovered that matrices, which he introduced for the description of atomic processes, did not commutate. It seemed to him that this could destroy his whole theory. But he forced himself to accept this result. Later this result was found not only fully correct, but also the most important one in the theory. Now we know that it was exactly this incommutability of the matrices that led to the famous «uncertainty principle» by Heisenberg.

It is well known how unwillingly M. Planck was forced to introduce the notion of quantum. For many years he attempted to get rid of it. But logical necessity proved to be stronger and the notion of quantum remained in science.

Thus in both logic and mathematics man created a means of self-compulsion which has become stronger than the creator itself.

* * *

We have considered in this chapter some moments of biological evolution — from the primitive control processes to the human consciousness, human creative ability, and intellect. These phenomena become the factors of further evolution — in the framework of Culture. Development of some cultural subsystems — like language and art — will be considered in the following chapters.

Chapter 3
Mechanisms of language

Perhaps, before one's lips was born a whisper, In a pre-forest air were whirling leaves...

Osip Mandelstam, Octaves.

It seems likely that during a definite stage of the evolution of *Homo sapiens* a need for communication among different human beings had to arise; this communication would serve to transmit current (non-genetic) information concerning the surroundings, joint labour, etc. The appropriate means of interpersonal communication did indeed arise and is known as *language*. We will not discuss the details of its physical («technical») realization (by means of acoustic, visual, or other devices), because as the basis of our consideration we take the appropriate needs and informational mental structures, which are to be involved in the processes language deals with; only on the basis of these processes it seems reasonable to construct the «technical means» that serve these processes. So, what are the assumptions that would permit us to deductively construct a system of language? (The logic of our further considerations about the structure of language is illustrated by Figure 3.1.)

One of the main ideas of the total theoretical construction on which we will base ourselves is the «evolutional» assumption that all the essential properties of information structures (such as language mechanisms) are nothing but the results of the constantly *growing efficiency* of information reception, storage, and processing. And an addition to this assumption is the so-called «*principle of succession*», i.e., we suppose that at a rather advanced stage of the evolution of mankind (at least at that stage when language had to appear), every new step («invention» of the evolution) had as its base the «shoulders» of the appropriate previous step, its most progressive achievements (devices, procedures, etc.). In other words, it seems impossible that a rather complex device could arise in an «empty place», without any predecessors. For instance, a wing of a bird could appear only on such a base as an arm; a system of colour vision became possible only on the base of monochrome photo-detectors, etc. So, this kind of «step-by-step» evolution can be seen as one of the most important conditions for the progress of information «technology».

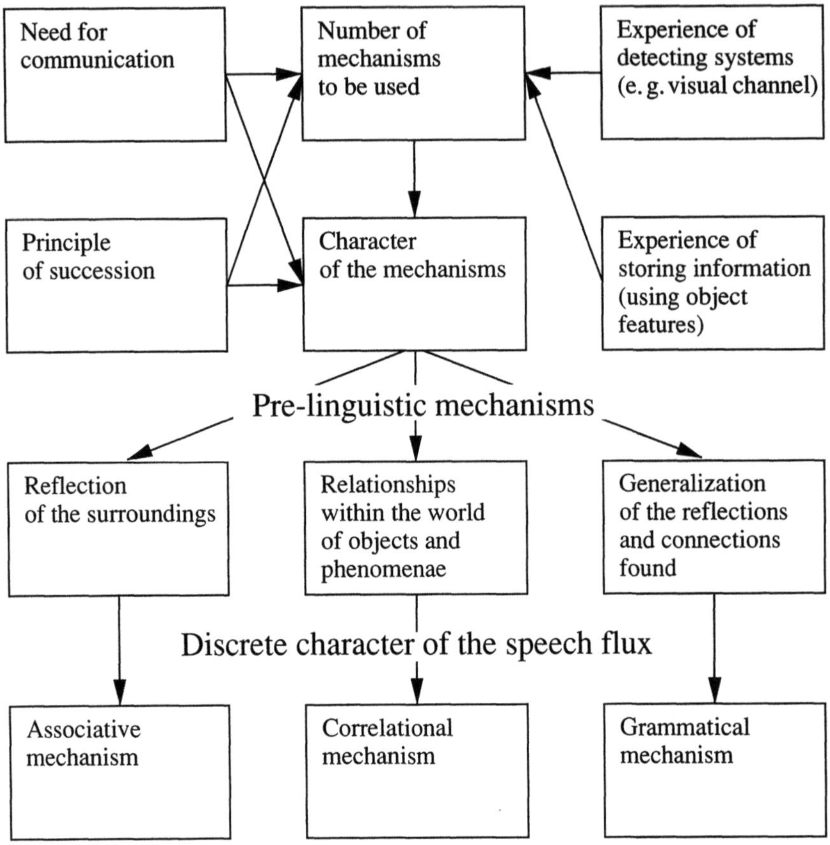

Fig. 3.1 Logical scheme of the deductive construction of language.

3.1 Language: How many mechanisms are involved?

If one accepts the above assumptions, it is necessary to find out what were the possible mechanisms that were used by evolution on its way to the «invention» of language. Two problems arise immediately when we start discussing this topic:

- How many mechanisms have to be taken into account, or what is the amount of pre-linguistic mechanisms that were involved in the birth of language?
- What could these involved mechanisms be?

To solve the first problem it seems reasonable to draw the most fundamental, general features of information processing. As to the amount of information units, the appropriate main feature is known as the advantage of the so-called «magic numbers» inherent to very different information processes. It seems that different devices and procedures dealing with information processing function most efficiently if they use a binary or ternary system of information encoding and transmission, i.e., if they are based on such *«magic numbers»* as 2 and 3 (see

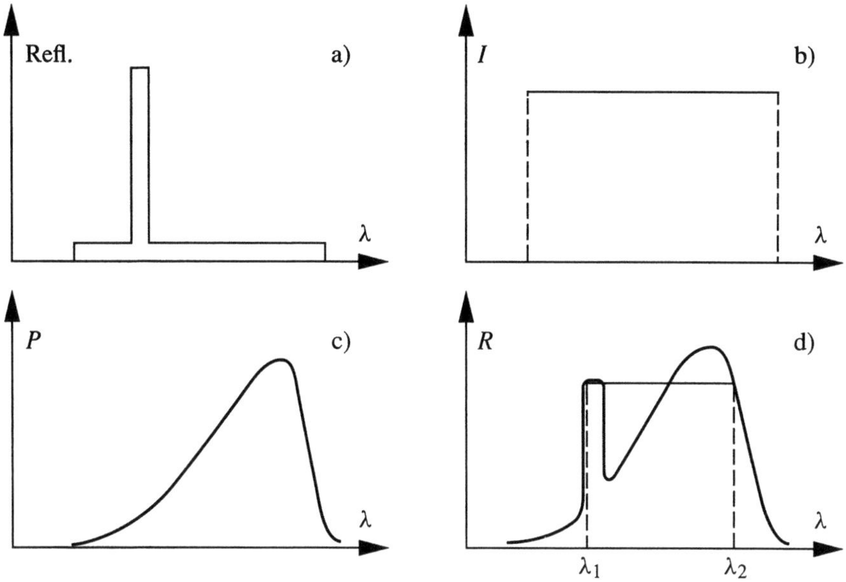

Fig. 3.2 The problem of identification of a property of an object perceived:
a) Reflection of the object's surface (Refl.) vs wavelenth (λ);
b) Intensity of illumination (I) vs wavelenth (λ);
c) Photoresponce of a detector (P) vs wavelenth (λ);
d) Ratio (R) of signals: from the object's surface and from the source of illumination – vs wavelenth (λ).

Chapter 1). This remarkable feature of information processing can be illustrated by many examples; here it seems enough to put forward only two of them.

1.) Let us consider the process of *identification* of an arbitrary *property* of objects, by means of a set of detecting devices. Let this property be a *colour* of an object perceived. (It seems reasonable here to choose the colour, because the identification of this property is one of the most «recent inventions» of evolution, and quite close to the stage of the birth of language.) We shall look at the simplest situation, where an object has a monochrome surface (when the colour of the object perceived is characterized by only one wavelength). It means that the curve for the reflection coefficient of an object is almost zero all over the visual spectrum, except for a very narrow fragment of this spectrum, where this coefficient is quite high (Figure 3.2a). How can we determine the meaning of this fragment (the wavelength responding to this narrow band), i.e., identify the colour of the object perceived?

(Such an identification can be very important when a subject (a human being or one of his evolutionary ancestors) uses colour for his practical needs. For instance, if an apple is green, it is not edible yet, but if an apple is red or pink, it is edible and, moreover, very tasty. So colour is a very useful property of the object perceived.)

Let us assume, as well, that the object is illuminated by a definite source of light, e.g., by a source having no spectral selectivity within the spectral range considered (Fig. 3.2b). Finally, our last assumption is that for colour perception, a subject (an organism) uses a detector with a bell-like spectral characteristic (Fig. 3.2c). Such spectral-selective behaviour of a photoresponse, i.e., a more or less sharp slope (increasing or decreasing the photoresponse with the wavelength) is absolutely necessary for our detecting purposes, because this slope provides us with an opportunity to identify the wavelength of maximal reflection. Besides, such a bell-like spectral characteristic is inherent to almost all types of light detectors, except for thermal ones, because any detecting device always uses some physical effects, each of them naturally having definite limitations. (For example, photoeffect in semiconductors usually has its only sharp maximum near the wavelength of the energy gap of the material used; that is why the photons responding to a longer wavelength cannot be detected, because of their too low energy, while at the shorter wavelengths the photoresponse is also very weak due to surface recombination processes.)

Being equipped with such a device, an organism (a subject) can try to identify the colour of an object perceived, i.e., the wavelength of maximum reflection of the object's surface. But it is impossible to do this by means of simple registration of the rays reflected from the surface of the object: the intensity of this reflected light depends on the intensity of the source of illumination. The only way to identify a colour is to compare the reflected light with the incident one, i.e., to use the ratio between the signal from the object surface and the «pattern» signal from the source of illumination (e.g., sunlight). Then the identification of a colour becomes possible, because this ratio is connected with the wavelength of a reflecting surface (Fig. 3.2d). Nevertheless, this connection is not simple: two different wavelengths (λ_1 and λ_2) respond to the same ratio of photoresponses ($\bar{R}$). It means that identification is possible only partly, not for all the wavelengths responding to the spectral range of our light detector.

How is it possible to improve such a situation? It is evident that one can try to use two detectors, their spectral characteristics being more or less shifted in relation to each other. The appropriate analysis (Petrov, 1984/Ref. 3) shows that in this case identification becomes more reliable, but still not for all the wavelengths. And only if one has a system of *three* detectors, their spectral characteristics being appropriately shifted — only then it becomes possible to determine (identify) each wavelength using the relationships of their responses. That is exactly why Nature invented 3-detector systems for colour identification, and a human being is equipped with one of these systems. So, 3-channel information transmission became one of the latest innovations of evolution. It is almost self-evident that our consideration can refer not only to colour vision, but to *any continuum*, where the identification of a definite object property is required; in all these cases it seems advantageous to use the *three-channel principle* of information processing.

2.) Our other example deals with the following problem: how to *store* information using the most economical means?

In general, there can be *two ways* to store information about different objects. The first way consists in simply keeping the *images* (or the names) of all the objects a given subject needs; these may be, e.g., the images of a hare, a tiger, an apple, a woman, etc. In this case the amount of primary elementary units (images or names) is equal to the number of objects which are needed for the subject's memory. (For instance, for 20 objects it is necessary to store 20 primary elementary units, i.e., images or names.)

The second way is to use some «*features*» of the objects, these features being the results of observations of different objects. For instance, such a feature, as «height», appears as a result of observations of very low objects, not so low objects, not so high objects, and very high objects; such a feature, as «tastiness», appears as a result of observations on different kinds of food, etc. These features can be measured on different scales — either continuous or discrete. Sometimes both types of scales can be used for measurement; e.g., the height of a tree can be measured using either a continuous scale (i.e., centimeters) or a discrete one (gradations: low, middle, or high); sometimes only a discrete scale is applicable; e.g., the sex of a subject has only two gradations (male or female). But a *discrete* scale is more universal than a continuous one, because all the features can be expressed with the help of a set of gradations, while continuous scales are capable of expressing only «ordered» features. Besides, to distinguish different objects, a subject needs borders between gradations of certain features, i.e., discrete scales. So in our further consideration we shall deal only with features measured using discrete scales.

There may be many features describing different aspects of objects, these features being distinguished (identified) by a subject. And it seems *more advantageous* to store information *using* these *features* (as primary elementary units), than simply the images of objects. For example, to distinguish 128 objects using their images, it is necessary to keep in memory 128 primary elements (images or names), while it is possible to use only 7 features, each having a discrete scale with 2 gradations, and there are possible $2^7 = 128$ different combinations, and it is necessary to keep only $7 * 2 = 14$ elementary units (gradations of features). So, in this case the second way of information storage is more efficient in $\frac{128}{14} \cong 9$ times. And when we go to more numerous sets of objects that have to be distinguished (identified), the advantage of the second way sharply increases. For instance, to distinguish 256 objects it is necessary to have 8 features, each with 2 gradations ($2^8 = 256$), i.e., to store only $8 * 2 = 16$ primary elementary units, instead of 256 images (or names) if one uses the first way; in this case the second way of information storage is $\frac{256}{16} = 16$ times more economical! That is why the second way had to be chosen by evolution and will be looked at further.

In the previous examples we dealt only with 2-gradation scales (i.e., each feature was described by 2 terms, or gradations). But are such scales really the most economical to keep information? What could be a set of features that permit the storage of information with the most efficiency? This problem was looked into by several investigators (Suchotin, 1983; Golitsyn and Petrov, 1990; see also

Chapter 1), and their results appeared to be quite identical: the most advantageous set of features must possess only 2- *and 3-gradation scales*. For instance, if it is necessary to distinguish 18 objects, it is possible to use:

- only two-gradation features; in this case it is necessary to have 5 such features (because $2^4 = 16$, i.e., less than a required number of 18, and $2^5 = 32$ is enough for this amount of objects); therefore, it is necessary to keep $5*2 = 10$ elementary units;
- only three-gradation features; 3 such features are required (because $3^2 = 9$, that is not enough for our amount of 18 objects, and only $3^3 = 27$ is enough); therefore, it is necessary to keep $3 * 3 = 9$ elementary units;
- a combination of 2 three-gradation features with 1 two-gradation feature (the resulting amount of different combinations is $3^2 * 2 = 18$, i.e., equal to the required amount of objects to be distinguished); here it is necessary to keep $2 * 3 + 1 * 2 = 8$ elementary units.

It is evident that the last variant is more advantageous than the two other variants, i.e., the combination of 2- and 3-gradation scales is the most economical.

This seems enough to illustrate the main point of this section: structures of information processing do really possess the «preferred» («magic») numbers 2 and 3. Evidently, these «preferred» numbers have to determine many specific properties of information structures inherent to all living systems, including the human one.

Among these specific properties are, for instance:

- the volume of the first step of the human memory, or the so-called «Magic Number 7, Plus or Minus 2», that was discovered by Miller (see Golitsyn and Petrov, 1990); this psychological phenomenon corresponds to a 3-cell operative memory device, each cell having 2 states (gradations); so a total amount of $2^3 = 8$ different combinations of states is possible;
- the threshold of perception, which is approximately constant for different stimuli (e.g., intensity of light, sound, electric current, etc.); for different stimuli the threshold is near 12–15%; this phenomenon can be explained (Golitsyn and Petrov, 1991) on the basis of the above-mentioned 3-cell operative memory device with 8 states: when all 8 states are occupied, to empty one of them means to lessen the stimulus intensity to $\frac{1}{8}$ of its magnitude, i.e., to 12.5%, and when only 7 states are occupied, to add one state means to increase the stimulus to $\frac{1}{7}$ of its magnitude, i.e., to 14.3%;
- the three-dimensional semantic space, that was discovered using semantic differential techniques (Osgood, Suci, and Tannenbaum, 1957: 290–304), as well as, maybe, also the apparent 3-dimensionality of our usual space.

So, binary and ternary structures penetrate all systems of information processing, especially in rather advanced stages of evolution. And *three components*: three gradations, three channels, three mechanisms, etc. should be the *upper limit* — the fundamental base for most of the information procedures, at least at that stage of evolution when language was to appear. So it seems reasonable to construct a model for language based on *three mechanisms* of information processing.

3.2 From pre-linguistic mechanisms — to language

Now our task is to find those three mechanisms of information processing that could be responsible for the emergence of speech communication. Returning to the succession principle, we have to consider some fundamental *pre-linguistic mechanisms* of information processing that could be modified to be used by language. What could these three mechanisms be?

N1) First of all, we have to take into account that the aim of speech communication is to carry some concrete facts about our surroundings. That is why *reflection of the surroundings* had to become the first mechanism that obviously participated in the birth of language. This mechanism (which in the case of human perception leads to sensuous images of objects perceived) is quite ancient, which can be proved by its localization in the right cerebral hemisphere (which is considered rather «archaic»).

N2) The next phase in the progress of information structures (or, in other terms, the next step of evolution) was the determination of the *relationships* within the world of objects and phenomenae. Naturally, this requires the ability to mentally divide the surroundings into discrete parts, i.e., to operate with discrete objects or phenomenae. Later on, this mechanism became the main reason for the discrete character of speech, which is divided into separate words. The same discrete character is also peculiar for the third (N3) mechanism described below, dealing with discrete grammatical categories.

The simplest form of such relationships is nothing but a *correlation* between definite objects: between the colour of an apple and its taste, between lightning in the sky and accompanying thunder, between a knock at the door and the arrival of a guest, between going to the forest and gathering mushrooms, etc. Some of such correlations were studied as so-called «conditioned reflexes» in the framework of a scientific school by academician Ivan Pavlov.

A propos, these very correlations, or connections, make dreams possible, i.e., a sequence of events where only some of the elements are interconnected with each other. It is known that in dreams time is usually inverted. This fact proves mainly that correlations function in the state of dreams, and not causal relationships. For instance, when a man is asleep, and his feet are touched by something hot — he could dream the following: he gets a phone call, is invited to Africa to his friend's house, goes there by plane and after that he walks through the Sahara desert, on very hot sand. Here we can observe an inverted sequence of events, ruled only by correlations: something hot (that touches a man's feet) is connected with hot sand, the last is connected with the Sahara desert, which, in turn, is connected with Europe by plane, a motive to go to Africa is connected with a friend living there, etc. So only connections — without causal relations — give a «degenerated» version of thinking realizing itself in a dream.

And, vice versa, the violation of such connections is typical for absurdist literature. For example, Eugene Ionesco uses such violations in his play *La cantatrice chauve*, where a knock at the door sometimes means the arrival of a guest, but sometimes not. In general, such a device — i.e., the violation of connections that became habitual for perception — is a basis for many phenomenae inherent to literature: humour, satire, the grotesque, etc.]

Such connections (correlations) are also — like the reflections of the surroundings (N1) — rather ancient; both the first mechanism and the second one are inherent to the pre-human step of evolution and are localized in the right cerebral hemisphere. They are absolutely necessary for the survival of any organism, and its successful functioning within its own environment. Both mechanisms operate with concrete sensuous images of objects to be reflected, and all the opportunities of concrete sensuous reflection of the surroundings are limited by these two mechanisms.

N3) Finally, the next step of evolution deals with another invention in the technology of information processing: the *generalization* of the connections found. For instance, a combination of several features: approximately round form, size of some centimeters, green or pink or red colour, etc. — leads to such a «concept» as «an apple»; then this combination (neglecting many other concrete properties of each given apple) is stored (in the subject's memory) as a whole — for different needs of the subject, e.g., for looking for food. So this way of keeping information seems to be quite useful; therefore, in the process of evolution it leads to a multilevel structure of information processing (Golitsyn and Petrov, 1990, 1991; see also Chapters 2 and 4). Within this structure, information is processed at different levels, each level receiving it from the previous one, the criteria of choice (concerning information to be transmitted to the next, higher level) being formed at each level. These criteria need «concepts» like those mentioned above, and the appropriate «concepts» serve as «instruments» in this kind of information processing. So, generalization is one of the most important mechanisms of information processing, and is, by the way, also inherent to the pre-human stage of evolution.

It seems natural to *combine* these three mechanisms of information processing (NN 1, 2, 3) in order to build a structure for communication, for language. Naturally, each of these mechanisms can be slightly modified to suit the needs of language; nevertheless, the main («immanent») traits of each mechanism must be maintained. Let us combine these three mechanisms with each other.

We have to consider a *flux of discrete signals* that can be used in a communication to carry information to a subject. (The discrete character of this flux, i.e., that it is divided into different words — is caused by the necessity to search for connections between the elements of this flux.) What can be the functions of these three mechanisms when a subject receives such a flux — a sequence of discrete speech signals?

We will have to use all the properties inherent to the above mechanisms, but all these properties have to be applied to certain specific objects, like fragments of speech communication.

While perceiving a speech flux (a sequence of discrete signals), the first (N1) mechanism can lead to the appearance of definite sensuous *images*, evoked by each discrete element of the flux (i.e., by different words) or by some of them together. For instance, an element «*table*» can evoke associations concerning an object with a flat surface and four legs. When applied to speech, this mechanism can be called the «*associative mechanism*».

This associative mechanism deals with certain discrete elements, i.e., segments of a speech flux, which are capable of carrying elements of *meaning* (or sense — here we do not distinguish these concepts) to a perceiver. These segments (fragments) can be, in principle, of rather different sizes; the minimal meaningful elements of a speech flux are well known in linguistics and called *morphemes* (see, e.g., Kuznetsova and Yefremova, 1986); sometimes a morpheme coincides with a word, sometimes it is a meaningful part of a word, e.g., its root, suffix, etc. Nevertheless, the mechanism considered — as well as other mechanisms which are described below — can operate not only with such minimal meaningful elements, but also with their combinations, e.g., *words* (and even with combinations of words). So to simplify our consideration we shall use the term «*words*», though quite the same mechanisms can sometimes function when using such segments, as morphemes, their combinations, and even combinations of words.

At the same time, the second (N2) mechanism can give rise to a special kind of *correlation*, like the *coexistence* of some definite fragments (words), i.e., the facts of their joint appearance in a previous speech experience of a given subject. For instance, this mechanism can check whether the following combination of signals: «*table*» and «*big*» has occurred in previous speech communications, or whether this combination is new for the given subject. This mechanism can be called the «*correlational mechanism*».

But when we receive a flux containing many different fragments (i.e., separate words) — *what pairs* of them have to be checked in such a way?

To answer this question one has to draw the third (N3) mechanism. Specializing in *generalization*, this mechanism can assign each fragment (word) to a definite abstract category, and among these categories there can be some stable links (interrelations). For instance, in Russian there are some grammatical rules connecting words to one another — *depending on* their *grammatical categories*: an adjective can occur with a given noun only if they are identical in relation to such categories as gender, number, and case. So, first of all a subject has to assign each fragment (i.e., each word) to the appropriate grammatical category, part of speech, etc. (In many languages it is necessary to take into account the word order within the sentence: e.g., the fact that a predicate always follows the subject.) Only after this, a subject can check the appropriate pairs, these pairs being determined by some grammatical rules. For instance, in a Russian sentence:

«Яувидел зеленыи стол» («*Ja uvidel zelyony stol*» — «*I saw a green table*») — a subject perceives concordance between words «зеленыи» («*zelyony*» — «*green*») and «стол» («*stol*» — «*table*») in gender, number, and case, as well as the concordance between words «увидел» («*uvidel*» — «*saw*») and «стол» («*stol*» — «*table*»), consisting in the transitive relationship between the verb and the noun. So, certain grammatical rules help a subject to look for the appropriate pairs of speech fragments (words) to be checked, and this is the function of the third (N3) mechanism during speech perception; that is why this mechanism can be called «*grammatical*».

In this *complex of mechanisms* only the second (N2) and the third (N3) one are «purely linguistic», specific for speech communication. The first (N1) mechanism is not exclusive for language; it is common for very different sign systems, because it deals simply with associations, inherent to separate signs (and in the case of a speech flux — with its fragments, i.e., words). As to the other two mechanisms (NN2 and 3), it is evident that only the *second* one (N2) is capable of carrying the so-called «*content*» of communication, because the third (N3) mechanism deals only with abstract categories and, moreover, serves only as auxiliary means for the functioning of the second mechanism. How can this mechanism present a «content» to a subject?

The second mechanism works in the *space of links*, connecting different words with each other. What could be the *structure of* this «*world of words*», i.e., the structure of this *network*? Let us consider this problem in the light of the reception of a new speech communication.

Each speech communication (if it is not simply a noise) contains some links between different words (indicated with the help of the third mechanism), that have to be *compared with the existing links* in the «world of words» of a subject — the recipient of this communication. It is evident that in this relation, *three situations* are possible in principle (see Figure 3.3):

a) A given pair of words (carried by this given communication) has been *connected earlier* in the previous speech experience of the subject. It means that the link between these two words carries no news, so this fragment corresponds to trivial communication. An example of a phrase containing such a pair of words is: «*Wadding is made of cotton*». Naturally, such a fragment can cause no specific emotion, connected with the content of this fragment. (An emotion in this case may be caused only by the *fact* of this trivial communication, this emotion being something like annoyance.) Due to the absence of a specific emotion, no special (speech) activity is provoked to deal with the re-building of a network of links within the subject's world of words.

b) Connection between words of the given pair has *not occurred earlier* (in the previous experience of the subject), but these two words *can be connected by* a certain *third word*, which has links with both of the two. This situation responds to the *emergence of closed chains* involving links among three or more words. This process can be interpreted as the perception of the so-called *sense* of

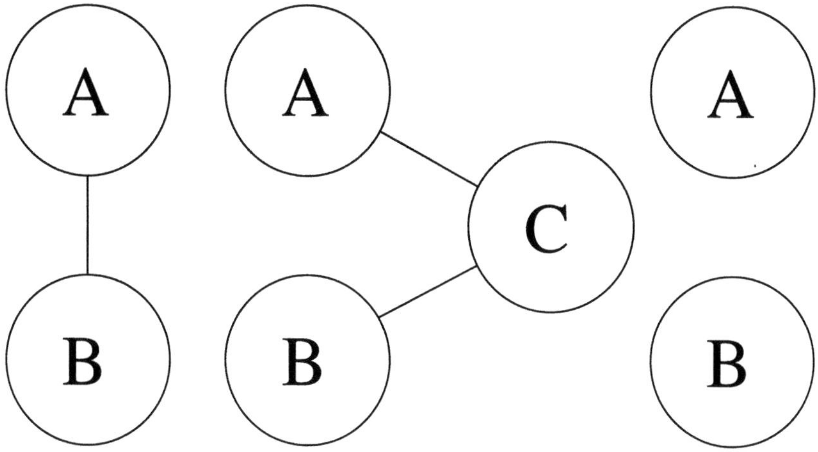

Fig. 3.3 Three variants of possible links between two perceived words A and B.

a communication. (It can lead to positive emotional reactions of the subject — due to the additional ordering of his mental structure and, hence, to the increase of the information contained within it.) So the *growth of knowledge* (received by means of verbal communication) — and, moreover, the *growth of* the entire *mental world* of a subject — is nothing else but the *emergence of new connections within* more and more *pairs of words*, together with a possible increase of the nomenclature of words perceived. (Some of the earliest connections may have been formed in a non-linguistic way, e.g., by visual associations, and afterwards some new connections are added to this network.)

The situation of the «formation of sense» can be illustrated by the following example. In the phrase: «*An oak is growing in the professor's garden*» — the statistical connection has to be checked between the words «*oak*» and «*garden*», as well as between «*oak*» and «*growing*»; and probably these pairs were already met together in the previous experience of the given subject. As to the words «*professor*» and «*garden*» — they may never have been met together. Nevertheless, they are linked via an absent (in the given phrase) word, «*man*» — because in the subject's experience there are links between «*professor*» and «*man*» and between «*man*» and «*garden*» (the relationship responding to the fact that a garden may belong to a man). But words of the pair «*professor*» and «*growing*» (as well as the words «*professor*» and «*oak*») were never met together in the subject's previous experience. So because of the known previous connections and the perception of the given phrase, some new links are forming, and all the main words used («*oak*», «*growing*», «*professor*», «*garden*») appear to belong to the same *network* (now because of the phrase above, the word «*professor*» becomes part of this network).

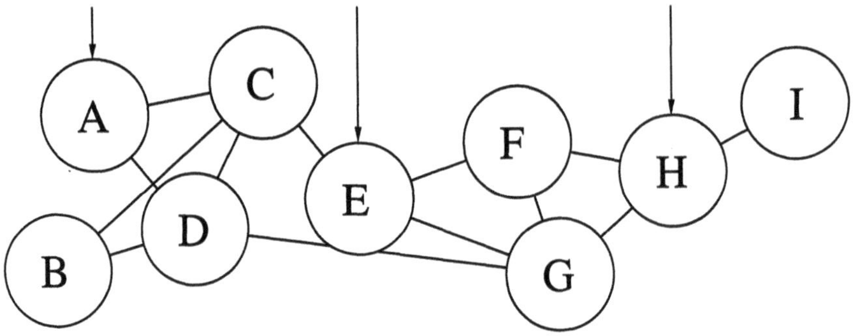

Fig. 3.4 Growth of the network of words' link as a result of perception of speech communication.

In this way, so-called «*sense*» is formed (as the expansion of the network — see Figure 3.4) during the perception of a series of speech signals — words.

c) Finally, it may happen that *no connection* within a given pair of words has occurred earlier, nor can the two words be connected by a third word. This situation corresponds to senseless (absurd) communication (eliciting a negative emotional reaction of the subject; the negative reactions force the subject to search his memory for connections).

Such are the results of our deductive construction of language, its mechanisms, and the formation of sense in a subject's mind, due to the reception of a verbal communication.

3.3 On language as one of the roots of literature

We have considered different versions of speech perception under conditions when the correlation mechanism and the grammatical one do function together; such functioning is specific for language communication. As for the associative mechanism, it can only colour the network above, by adding to its elements some sensuous components or even images. Besides, the associative mechanism can prove the rightness of the results of the complex process of speech perception. If a definite communication results, for instance, in a tight connection between the words «*hummingbird*» and «*iceberg*», — then the associative mechanism gives help signaling the mistake.

The main function of the joint functioning of the above-mentioned three mechanisms (NN1, 2, and 3) is to provide a subject with a sense of verbal communication. In addition to receiving this sense, different other relations of the results of the operations of these three mechanisms lead to the emergence of some interesting nuances of perception. For instance, such a specific phenomenon as *humour* is the result of comparing the processes of different mechanisms, e.g., mechanism N1 connects words A and B, but mechanism N3 forces one to connect words B and C, etc.

Such is the functioning of the mental structures during speech perception in a «normal» regime. But can this normal regime always work automatically, without failures? Or does it, maybe, need a certain special «*training*», through some «special communications» (as it happens in relation to many other kinds of information activity — see Chapter 4, as well as Golitsyn and Petrov, 1991; Petrov, 1993)? We have to suppose that there may exist at least several classes of objects that can be used for the training of the joint, coordinated functioning of the above three mechanisms. We shall not look at all of them, but only at the ones realizing themselves in *poetical texts.*

Why do we need to train the human information system in the joint functioning of these mechanisms? The reason is that this joint functioning is rather *complex* and, therefore, can be exposed to some dangers. Thus, one of the mechanisms can be reduced, weak, and even cease to work (especially if the mechanism had not been operative for a long time). Or maybe one of the mechanisms can work autonomously, without any coordination with other mechanisms (e.g., if, before this, it was the only really working mechanism, and the two other mechanisms were almost «sleeping»). As a result, we can observe different *violations* of information processing, possibly related not only to speech information, but also to all kinds of information that is kept in the long-range memory of a subject. In particular, a subject will more passively search for connections between words, and therefore will have a worse understanding of the sense of communications, will not so intensively generate new ideas, will lose part of his/her creative potential (will lose the capability for the vivid perception of ideas and impressions — i.e., all that the great Russian poet Alexander Pushkin named «inspiration»), etc.

To avoid this, it seems desirable to use some *special speech communications.* These communications have to succeed in putting into *active work all three speech mechanisms.* In turn, this requires that the *structure* of such special objects (communications) should be *organized* in an appropriate manner on the *levels of speech processes*, i.e., the distances should have an order of magnitude of *several words.* Generally speaking, the size of such special objects can be quite arbitrary, but they must possess a definite «short-range order» (in terms of crystallography). But we do know certain objects that consist of words and have an «ordering» on «short distances» — these objects are *poetical works* (sonnets, poems, ballads, etc.). All of them possess a rather definite organization on the scales close to a word; such an organization is facilitated by a *division* of the text into appropriate *small sets* (e.g., into lines and stanzas), metre, rhyme, etc.

We have developed something like a short *deductive construction* of the psychological *need in poetry* (and, at least, of some general features of objects capable of satisfying this need), starting from rather abstract considerations mainly drawn from information theory. We will not, however, realize an analogous construction of other classes of «special objects» which use speech mechanisms, like the different genres of *prose*; they are the object of a separate study (Petrov, 1993; see also Chapter 4). Moreover, here we cannot construct different poetical genres nor the structural features of the appropriate poetical works. In the framework of this book, we shall deal only with one aspect of the psychic processes accompanying the perception of poetical works (see Chapter 5).

3.4 The sound structure of language: Choice of the information channel

One of the most interesting results of our investigations was the necessity to divide a speech flux into *discrete signals* — words or morphemes (links among these discrete parts of a speech flux forming the so-called «sense» of a message). But we did not deduce the process of information *transmitting*, the nature of the signals which carry speech information, etc. Now it is time to solve these problems, i.e., to deduce the *sound structure* of language, meaning the general properties inherent to different natural languages. (The logic of our further consideration is illustrated by Figure 3.5; for details see also Petrov, 1984/Ref. 2; Golitsyn and Petrov, 1991.)

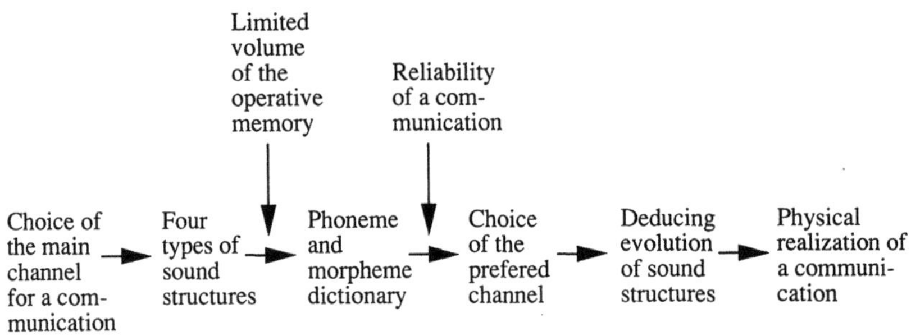

Fig. 3.5 Logical scheme of deducing the sound structure of language and its evolution.

It is well known that the main channel of speech communication (characteristic for languages of different peoples all over the world) is the acoustic one (e.g., Panov, 1983). But why is this so? Why does no natural language use — as its main means — the visual channel, the sense of smell, etc.? Let us deduce the choice of the main channel, accepting the assumption that at the initial stage of language development the human system of information processing was more or less identical to the system man possesses nowadays. (Of course, the development of language itself could cause some changes in the human information system, but not concerning its most general properties.)

The last assumption leads to the following result: only two channels can claim the role of prefered means of speech communication, these channels being the *visual* and the *acoustic* one. They permit information to be received from rather large distances, whereas all other information channels a human possesses (e.g., a sense of touch or a sense of smell) are suitable only for non-distant communication or for very short distances.

To choose the appropriate information channel, we have to compare the opportunities of each means of distant communication with each other. Which is the best way to transmit information from one person to another, or from a *transmitter* to a *receiver*? To answer this question, it seems reasonable to take into account such parameters of information channels as:

α) the *conditions* that make a communication possible;

β) the *quality* of a communication.

In turn, the conditions of a communication can be classified as follows:

$\alpha 1$) the *energy resources* needed for information transmitting; these resources can be divided into two parts:

$\alpha 1'$) the energy needed to generate signals (and to send them via a channel);

$\alpha 1''$) the energy needed to receive signals (and to amplify them);

$\alpha 2$) the *spatial conditions* of a communication, including:

$\alpha 2'$) the distance a communication has to cover between a transmitter and a receiver (e.g., several meters, several hundred meters, or several kilometers);

$\alpha 2''$) the «configuration» of a communication, characterizing the properties of a medium through which this communication can be realized (e.g., transparency of this medium for electromagnetic waves or acoustic ones);

$\alpha 3$) the *number of people* that can simultaneously participate in a communication (both as transmitters and receivers).

As to the quality of a communication, it can be considered to consist of:

$\beta 1$) *time resources* needed for information transmitting; these resources can be divided into two parts:

$\beta 1'$) time needed to generate signals;

$\beta 1''$) time needed to cover a distance between a transmitter and a receiver;

$\beta 1'''$) time needed to receive signals;

$\beta 2$) *reliability* of information transmitting, including:

$\beta 2'$) reliability of generating signals (meaning possible distortions while encoding information);

$\beta 2''$) reliability of covering a medium (meaning also possible distortions) between a transmitter and a receiver;

$\beta 2'''$) reliability of receiving signals (meaning possible distortions while decoding information).

So, we have at least 11 parameters ($\alpha 1' - \beta 2'''$) with which to compare our two information channels (visual and acoustic). Here we have to imagine the more or less definite obstacles that caused a need for speech communication to arise. We may suppose that this communication arose during a joint activity of different persons, who found distances of several meters (or, maybe, not more than 100–200 meters) between them; such distances are typical, e.g., for hunting, fishing, etc. In these cases, what are the advantages of the visual channel and of the acoustic one?

It appears that in respect to these obstacles, the overwhelming majority of the above parameters display arguments neither for the visual channel nor for the acoustic one. Really, the energy needed to generate signals ($\alpha 1'$) is, evidently, approximately equal for both channels, because it is necessary to put some parts of a human body in motion — e.g., a tongue — in order to give the signal with the help of voice, an eyelid — to signal by winking, an arm or a finger — to signal either through visual signs or through acoustic ones (meaning knocking); it is impossible to a priori prefer one of the above channels in relation to this parameter ($\alpha 1'$). Most of the other parameters are in a quite analogous situation (we shall not tire our readers with appropriate considerations), except the «configuration» of a communication ($\alpha 2''$) and the number of persons participating in a communication ($\alpha 3$). Here the acoustic channel possesses some important advantages.

First, the *«configurative» conditions* of a communication ($\alpha 2''$) are substantially better when using the acoustic channel, because it deals with longitudinal waves — oscillations of the density of the air, with wavelengths of several meters; that is why these waves can skirt many obstacles in their path. And the visual channel deals with electromagnetic waves with wavelengths less than 1 micron, which cannot skirt different obstacles in their path. Therefore, the acoustic channel has many advantages when working in a non-homogeneous medium, e.g., in a forest (consisting of trees which are not transparent for electromagnetic waves). Second, the *number of people* that can participate simultaneously in a communication ($\alpha 3$) is much more when using the acoustic channel, because the visual channel makes it necessary to focus an eye mainly on one object (i.e., a subject that transmits information), whereas the acoustic channel permits information to be received simultaneously from many transmitters, e.g., from several subjects hunting in a forest.

So though the merits of both channels are almost equal, the *acoustic channel* had to appear slightly more advantageous in the circumstances accompanying the origin of language. This was probably the principal reason for the acoustic channel to be chosen — as the main means of a speech communication — by all the peoples all over the world. Nevertheless, under certain circumstances the visual channel can compete with the acoustic one; therefore, some primitive societies had developed languages based on the visual channel (primarily destined for communication at very long distances). But the acoustic channel was chosen as the main line of speech communication, and it is this method that is the object of our further consideration.

Let us consider the possible ways to realize speech communication on the basis of the acoustic channel. As mentioned above, while perceiving a speech flux, a subject has to combine different segments of this flux with each other in order to receive the so-called «sense» of the communication. Therefore, a subject must *identify* at least the *minimal segments* capable of carrying elements of sense; these elements are nothing else but *morphemes*. (Sometimes a morpheme coincides with a word, sometimes it is a meaningful part of a word, e.g., its root, suffix, etc.) So our first task is to find ways of organizing a speech flux that permit a receiver to *distinguish* different *segments*, i.e., to *identify* different *morphemes*.

3.5 Phonetic structures: Possible ideal types

As soon as our segments are of an acoustic nature, it seems reasonable to construct them taking into account two of the most important aspects:

a) the temporal behaviour of a segment (i.e., its behaviour in time);

b) the qualitative characteristics of its phonation at each given moment.

Evidently, the first aspect (a) is the most important of these two, because temporal division is the main condition for separating a speech flux, for finding separate segments (morphemes) within it. So we must analyze *behaviour in time* from the most fundamental point of view, in relation to the process of perception: we can classify all possible segments depending on whether their perception is characterized by *integrity* or not; as a result two «polar» types of segments are possible:

a1) each segment (morpheme) is perceived *as a whole*, a temporal integrity, i.e., *without being divided in time* into separate parts;

a2) each segment is perceived *as a temporal sequence of* some *components.*

Turning to the second aspect (b), we can classify all possible segments on the base of the same fundamental ground, which was used above: the integrity or non-integrity of perception; here two «polar» types of segments are also possible:

b1) each object (a segment or its component) is perceived *«synthetically», as an integrity, without distinguishing* any specific *features*, which could be used to compare different objects (segments or their components) with each other;

b2) each object is perceived *«analytically», as a carrier of* certain *features*, which can be used to distinguish it from other objects.

The above two classifications, combined with each other, yield *four possible «ideal» types of sound structures* (a1b1, a1b2, a2b1, and a2b2), with an evidently different nomenclature of segments, which can be used during speech perception as primary meaningful units (morphemes), identified by a perceiver. Let us consider these four «ideal» types, stressing the nomenclature of segments, which can be used by each of these types. Naturally, each of these four types is an «ideal» one, a pure realization of the main classificational ideas; in reality (and especially during evolution — see below) these types can be mixed; nevertheless, it is quite useful to take a closer look at them.

1. In the **first type of structure** (a1b1), the perceived segments themselves constitute the primary (elementary) units. Such a structure can be realized in a set

of definite *«standardized» sounds*, each of them being an *individual integrity*, i.e., not divided temporally (on certain parts) nor qualitatively (on certain features). Therefore, in this case each segment (morpheme) consists of the only «standardized» acoustic signal, which is called *«phoneme»* in linguistics. For instance, such morphemes (phonemes) may be sounds «E», «O», etc. (It is interesting to note that a Russian poet, Velemir Khlebnikov, wrote: «Primitive tribes were inclined to use names consisting of one vowel only» (Khlebnikov, 1986; p. 204); that is why he entitled one of his best poems: «E and A. A Tale of the Stone Age».) How extensive could a nomenclature of such sound structures be?

Here we must take into account that the perception of a speech communication demands a rather *high speed.* (This requirement is applicable to each type of sound structures though it will not be repeated below.) This means that each segment (a morpheme, and in the given case — a phoneme) must be recognized (identified) almost instantly, i.e., by means of the first step of memory — the so-called *«operative memory»*. But it is known (and this fact was deduced from the general information model — see, e.g., Chapter 4), that this step of memory has a very limited capacity: it is capable of storing not more than eight elementary units (images or other objects, each of them being perceived as a certain integrity). Therefore, an amount of *eight phonemes* is the upper limit for a nomenclature of phonemes of such structures (and, naturally, the nomenclature of morphemes coincides with this phoneme nomenclature).

2. The **second type of structure** (a1b2) deals with segments, each of them still temporally undivisible (as in the previous case), but capable of being analyzed, i.e., singling out definite specific *features*. This situation can be also realized in a set of *«standardized» sounds*, each of them responding to a separate morpheme. But here the role of elementary units is played not by these «standardized» sounds themselves, but by their *specific features*. Therefore, here the amount of different phonemes (again coinciding with morphemes) is much higher than in the previous version. To estimate this amount, we have to take into account the above-mentioned limited capacity of the operative memory (eight units), and apply this restriction to the features considered. What information can be kept in the operative memory in this case? Here two variants are possible, corresponding to features that have a continuous scale or a discrete one.

Let us assume that a feature has a continuous scale, but this scale can be divided into several gradations (for the aim of using such division to distinguish phonemes). For instance, such a feature, as a *«duration»* of phonation, can have three gradations: short sounding, middle sounding, and long sounding ($D_s - D_m - D_l$). This means that it is necessary to keep (in the operative memory of a recipient) the values of borders between the adjacent gradations (i.e., between D_s and D_m, as well as between D_m and D_l). In general, if a feature with a continuous scale has x gradations, we must occupy $x - 1$ cells of the operative memory to keep information about these gradations. Then let us assume that a feature has a discrete scale with x gradations; in this case it seems reasonable to keep (in a recipient's operative memory) also $x-1$ meanings, i.e., all the gradations themselves, except one of them (because if a signal received does not respond to these kept gradations, it evidently responds to the only remained unkept gradation). So, both variants are practically equal: to work with a feature which has x gradations, it is necessary to occupy $x-1$ cells of a recipient's operative memory. But if we want to use more than one feature — how can we distribute the cells among these features?

For example, we want to use two features: «duration» and «loudness». It is quite easy to prove that the most advantageous distribution of cells responds to equal numbers of cells for each feature. So, if four cells are occupied by «duration» and four cells by «loudness», then we can have five gradations of «duration» ($D_1 - D_2 - D_3 - D_4 - D_5$) and 5 of «loudness» ($L_1 - L_2 - L_3 - L_4 - L_5$). So the total amount of different combinations of these two features ($D_1L_1, D_1L_2, D_1L_3, \ldots D_5L_5$) is $5 * 5 = 25$. Such is the nomenclature of morphemes (which coincide with phonemes) in this version. If we want to use four features, we can have a nomenclature of 81 morphemes (phonemes), etc. It can be easily shown that the *most advantageous* version responds to eight features, information about each of them occupying one cell — therefore, each feature has only *two gradations*; in this case the nomenclature of different segments will be quite rich: $2^8 = 256$ *morphemes (phonemes)*!

The last conclusion is important for the functioning of language (as well as for its development): it is clear that if this type (a1b2) of sound structure is realized, it must operate with *2-gradation features*! Moreover, the *2-gradation principle of information storage* appears to be essential not only for language mechanisms, but also for a wide circle of other phenomenae, based on the properties of the operative memory. Maybe the most important manifestation of this principle is the dominance of 2-gradation features (which are sometimes called dichotomic ones, or binary oppositions) in *anthropology, cultural sciences and related fields*, and this fact has become the basis for some models of contemporary structuralism.

3. The **third type of structure** (a2b1) creates segments (morphemes) by building certain chains (sequences) consisting of *temporal components*, each of them being an integrity, that is perceived as a whole. So, here the role of primary elements perceived is played by the phonemes themselves. What is the best way to distribute the cells of the operative memory among these phonemes?

Let a segment consist of two such «holistic» phonemes: X and Y. In this case we have to assign to the first phoneme x cells and to the second phoneme — $y = 8 - x$ cells, so a nomenclature of X-phonemes is x, a nomenclature of Y-phonemes is $8 - x$, and the resulting nomenclature of different 2-phoneme segments will be $xy = x(8 - x)$ units. This value is maximal when $x = 4$, and the total amount of 4(8 - 4) = 16 different segments (morphemes) is possible. (If the nomenclature of the first phonemes is X_1, X_2, X_3, X_4, and the nomenclature of the second phonemes is Y_1, Y_2, Y_3, Y_4, then these 16 morphemes are the following: $X_1Y_1, X_1Y_2, X_1Y_3, \ldots X_4Y_4$.) It was proven that if a segment consists of three phonemes (X, Y, and Z), then it is most advantageous to assign three cells to one of them, to another one — also three cells, and to the third one — two cells; so the total amount of different segments (morphemes) is $3 * 3 * 2 = 18$ units (the examples of such morphemes: $X_1Y_1Z_1, X_1Y_1Z_2, \ldots X_3Y_3Z_2$). If a segment consists of 4 phonemes, then the nomenclature of different segments contains 16 units; 5-phoneme segments have a nomenclature of 8 units; 6 phonemes are capable of forming only 4 different segments, 7 phonemes — only 2 segments, and 8 phonemes — only 1 segment (morpheme).

So if this type (a2b1) of sound structure is realized, it must use mainly *2, 3, or 4 phonemes*, and the nomenclature of morphemes will have 16, 18, or 16 units respectively. As to the nomenclature of phonemes, it consists (in all the cases) of *8 units*. (Thus, 3-phoneme morphemes can be constructed of the following phonemes: $X_1, X_2, X_3, Y_1, Y_2, Y_3, Z_1, Z_2$.)

4. Finally, in the **fourth type of structure** (a2b2) we deal with *temporal components*, and the role of *primary elements* is played by certain *features*, i.e., each temporal component is perceived not as an integrity, but as a *carrier of* a set of definite *features* (which are called «differential» ones in theoretical phonetics).

Let us assume that in respect to gradations we realize the most advantageous version, i.e., we use only 2-gradation features. Then let these features relate to 2 phonemes: x features to the first phoneme, and $8 - x$ features to the second phoneme; then the nomenclature of the first phonemes consists of 2^x *units*, and of the second phonemes — 2^{8-x} *units*. (For instance, the first phoneme is described by 3 features, and the second one — by 5 features, i.e., information relating to the first phoneme occupies 3 cells, and to the second one — 5 cells; so the nomenclature of first phonemes will number $2^3 = 8$ units, and of second ones — $2^5 = 32$ units.) Using such phonemes, we can build $2^x * 2^{8-x} = 2^8 = 256$ different segments (morphemes). (In our example this morpheme nomenclature is $8 * 32 = 256$ units.) As to the nomenclature of different phonemes, it depends on the distribution of cells among the phonemes used (in our example this phoneme nomenclature numbers $8 + 32 = 40$ units). It can be proven that the phoneme nomenclature is minimal (i.e., the «stingiest») when we appoint to one phoneme 3 cells, and to another one — 5 cells (just as in our example), and then a phoneme nomenclature consists of $2^3 + 2^5 = 40$ units. And the maximal (the «richest») phoneme nomenclature arises if one phoneme is described by one feature, and another one by 7 features: then the phoneme nomenclature numbers $2^1 + 2^7 = 130$ units. But we have to remember that in all these cases, the upper limit of the morpheme nomenclature is constant and equal to 256 units.

In order to illustrate the realization of the morphemes in such sound structures, let us suppose that the first phoneme is described by 3 differential features: duration (D), loudness (L), and pitch (P), and the second one by 5 differential features: sharpness (S), consonantness (C), brokenness (B), flatness (F), and again loudness (L); and let each of these features have two gradations. Then the following different segments (morphemes) are possible:

$$- (D_1L_1P_1)(S_1C_1B_1F_1L_1);$$
$$- (D_1L_1P_1)(S_1C_1B_1F_1L_2);$$
$$- (D_1L_1P_1)(S_1C_1B_1F_2L_1);$$
$$\dots\dots\dots\dots$$
$$- (D_2L_2P_2)(S_2C_2B_2F_2L_2).$$

Turning to other variants (consisting of more than 2 phonemes), we shall present only the results of the appropriate calculations. Thus, when using 3 phonemes, the minimal nomenclature contains 20 units, and the maximal one — 68 units. When building morphemes of 4 phonemes, the minimal and the maximal phoneme nomenclatures consist of 14 and 38 units respectively, etc. And, at last, when using 8 phonemes, the nomenclature (which is minimal and maximal at the same time) contains 16 units.

As a result of this construction, all the versions of this type (a2b2) show a rather broad spectrum of phoneme nomenclatures: *from 14 units* (in 4-phoneme and in 7-phoneme morphemes) — *to 130 units* (in 2-phoneme morphemes). Nevertheless, the upper limit of all their morpheme nomenclatures is quite the same: *256 units*. Therefore, a very important conclusion can be drawn from our considerations: all sound structures of this type can be based on a varied phoneme nomenclature, made up from 14 to 130 units — but *all* these *sound structures* have *equal practical opportunities* (at least in respect to their morpheme nomenclatures) The main results of the previous considerations are summarized in Table 3.1.

3.6 Possible evolution of sound structures

So, we have at our disposal 4 types of «ideal» sound structures (and, evidently, a variety of their combinations) which, in principle, are capable of realizing verbal communication. Now it is time to attempt to construct the evolution of this means of communication, i.e., to build (using a deductive method) a *«trajectory» of phonetic evolution* that possibly took place during the development of human languages. Naturally, this trajectory must include the *starting point* of the evolution, as well as the *«final point(s)»*, which we can observe now in various natural languages all over the world.

The main principle of our construction will be a constant trend towards a more and more *perfect functioning* of the system considered. It means that we must construct — step by step — a path from the simplest sound structures, with very limited opportunities (in respect to carrying sense, which needs to have a definite nomenclature of morphemes) — to more complicated sound structures, with «rich» opportunities. As to the *simplest structures*, which should correspond to the *starting point* of our phonetic evolution, we have undoubtedly *only one «candidate»* for this role — it is our *first type* (a1b1), which needs almost nothing

N	Type	Temporal integrity	Qualitative integrity	Preferable phoneme nomenclature	Maximal morpheme nomenclature	Comments
1	a1b1	+	+	8	8	Morphemes coincide with phonemes
2	a1b2	+	–	256	256	Morphemes coincide with phonemes
3	a2b1	–	+	8	18	Each morpheme consists of 2, 3, or 4 phonemes
4	a2b2	–	–	From 14 to 130	256	Each morpheme consists of 1, 2, 3, 4, 5, 6, 7 or 8 phonemes

Table 3.1 General characteristic of the four «ideal» sound structures

for carrying out special operations, because signs without temporal division and without singled out distinctive features are evidently common for humans in all stages of their evolution. But as for more *developed structures* (possessing rather «rich» opportunities in respect to their morpheme nomenclature), which should respond to the *final point*(s) of our phonetic evolution — here we have *two «candidates»* possibly equally strong, i.e., *the second* (a1b2) *type* and the fourth (a2b2) *one*; so first of all we must try to *choose between* these *two types*.

Both types have a quite identical nomenclature of different morphemes (256 units); but maybe they differ in some other way? Here it seems reasonable to remember the parameters of communication ($\alpha 1' - \beta 2'''$) we dealt with when choosing the main information channel. Let us compare the two sound structures (a1b2 and a2b2) on these parameters. It can be easily proven that these two types have practically equal rights when using criteria concerning all the parameters, except one — the *reliability of receiving signals* ($\beta 2'''$). On this parameter the fourth type of sound structure (a2b2) is much better than its «rival» (a1b2), because of the following.

The reliability of receiving speech signals (meaning the identification of segments) is connected to the obstacle that different morphemes should not be too easily *confused* with each other. Of course, each of them differs from the other in at least one differential feature, but in cases of pure audibility and/or high level of noise this difference can not be pronounced enough for normal communication. For instance, the words «*FULL*» and «*FOOL*» have quite different meanings, but differ only by one differential feature (pronunciation of the vowel), and therefore sometimes a confusion may take place. It does not matter if such a confusion will appear only with several words (or morphemes): usually a context helps, i.e., a semantic redundancy (inherent, as a rule, to any speech communication) comes to the aid. But if a speech flux has many words (morphemes) like this? It could be dangerous, so it seems reasonable to increase the reliability of verbal communication. There are *two ways* to do this.

First, in order not to confuse different morphemes (or, at least, different words), it is sometimes desirable to fix a *temporal distance* within a speech flux, i.e., a *pause*. For this purpose one can reserve one cell of a receiver's operative memory; this cell will fix the presence (or absence) of an interval between neighbouring segments. Naturally, this will result in a decrease of the amount of the cells which are responsible for identifying phonemes themselves: now only 7 cells can be used for this. Both sound structures considered (a1b2 and a2b2) can take this course, and if it were to be realized to the full extent — then each type would have a quite *identical*, reduced *morpheme nomenclature* of $2^7 = 128$ units. But this course is not necessary: it can be realized or not, or realized only partially.

Second, it seems reasonable to require that either phonemes or morphemes should differ from one another not by one differential feature, but by two or even three features; in this case they most certainly will not be confused. (Really, as mentioned by Jakobson and Waugh, 1987; p. 28, «...the difference between a single and a multiple distinction $\langle\ldots\rangle$ is a palpable operation both for linguistic analysis and for actual interlocutors. Lew Tolstoj acknowledged that he endowed some characters of *War and Peace* with generally known aristocratic Russian family names, which he changed only slightly, so that they would sound familiar to his readership. In each case he played with one single feature, a letter, distinction; he turned *Volkonskij* into *Bolkonskij* and *Trubeskoj* into *Drubeskoj*. These names sound quite natural, whereas with a multiple alteration they would have become grotesque, for instance **Dolkonskij* or **Polkonskij* or **Zrubeckoj*».)

What can be the consequences of this (i.e., increasing the reliability because of phonetic redundancy) for a morpheme nomenclature in both types of sound structures?

It appears that, naturally, the volume of the morpheme dictionary will decrease, but this effect will be a lot less strong in structures of the fourth type (a2b2). For example, we want morphemes to differ in three features or even more. In this case a structure of the second type (a1b2) will possess only 15 morphemes, instead of 128 morphemes without such a «triple» requirement. And the same «triple» requirement being applied to structures of the fourth type (a2b2) comes up with a not unrespectable result: different variants are possible (depending on a phonemic structure of morphemes), but sometimes the reliability above can be attained even without any reduction in the morpheme nomenclature. For instance, if a morpheme consists of 3 phonemes, and the same requirement of triple distinction is applied, a nomenclature of different morphemes (when using features that are different for each kind of phonemes) is still $2^2 * 2^2 * 2^3 = 128$ units. So, when both structures (a1b2 and a2b2) have a morpheme dictionary of the same volume, one of them (a2b2) possesses a much higher reliability. (Naturally, this is caused by a rather deep «specialization» of the cells within the operative memory and, respectively, the appropriate, conditional «specialization» of the phonemes in the language used: each cell is allotted a definite kind of phonemes, so their sequence, or *order*, carries «additional» information. In general, all phonetic «progress» is connected with deeper and deeper specialization of the elements used. Thus, to increase the reliability, each language can also use some «stable», «standard» combinations of phonemes, responding to quite definite fixed positions in a word; but this «device» relates to a higher level of language functioning than the level of feature structures, and therefore will not be considered here.)

The main result of the calculations above is that the *fourth type* (a2b2) of sound structures *is* undoubtedly *more preferable* than the second type (a1b2). It means that the *development* of sound structures should be directed *towards the fourth type*, which has to be the *«final point»* of phonetic evolution. Let us try to reconstruct the trajectory of this evolution.

As the *starting point* of the evolutionary process (concerning sound structures used by a language) we can accept the *first type* (a1b1): of all the four types, this type contains the simplest signs, which need no special innovations (assuming human beings were already accustomed to usual simple signs at the stage when the sound structures began their development). But what were the next stages of the phonetic evolution, which had to be accompanied by some innovations?

Two main *«inventions»* are virtually working within our classification. The first invention consists in the replacement of an «instant» signal (in structures a1b1 and a1b2) by a temporal sequence of components. The essence of the second invention is to replace a synthetic, «gestalt» perception (in structures a1b1 and a2b1) by an analytical process, which can single out certain features in the sounds perceived. Both inventions appear to be «perfecting» means directed towards the creation of the structures on the basis of the first type (a1b1), though using different devices. Therefore, it seems possible to reconstruct the evolution process as a *parallel introduction* of these two inventions in the practice of sound structures, i.e., the parallel replacement of elements of the first type both by temporal-expanded elements and by «featured» ones. (And the relationship between rates of these two processes is not so important for our consideration, which deals primarily with the most general principles of phonetic evolution.) As a result, the *evolutionary trajectory* can be represented by the following *steps*.

During the *first step*, the simplest *«single signals»* arise, i.e., some structures of the a1b1 type. These are some «exclamations» («*E*», «*A*», etc.), which are used, e.g., by hunters to signal that a beast is coming, a given hunter is killed, etc. But the nomenclature of such signs is rather poor (not more than 8 units), and a need to increase it arises, i.e., for some modifications of the existing sound structures.

That is why the *second step* of evolution arises, consisting of *distinguishing* certain (maybe, individual?) «*hues*», in such «single signals», e.g., such features as a duration of phonation, its pitch, etc. This means that each single signal is now perceived as a *carrier of* several *features* that responds to structures of the second type (a1b2). For instance, a signal «*E*» may be perceived as: a long vowel, diffused, voiced, etc. Due to such an invention (innovation), the nomenclature of the morpheme dictionary greatly increases.

Parallel with this process (or, maybe, partly on the basis of some of its achievements), the *third step* of evolution comes into being. It seems possible to perceive some «hues», or features *in the duration* of phonation of each sound, e.g., its soaring up or down. It is nothing else but the beginning of the third type (a2b1) of structure, which has a «richer» dictionary than the first type. Due to this, such an innovation is consolidated; but it is much more important that together with this innovation a *device* is also consolidated, this device realizing the innovation.

This device, responding to the *fourth step of evolution*, consists in using mostly *temporal blocks*, each of them containing *two or three phonemes*, ordered in time. But using such blocks requires that special *positions* (in a temporal sequence) should be allotted to *each kind of phonemes*. So, a need appears to *divide* all the *phonemes* at least *into two kinds* (in order to ascribe each phoneme to the appropriate temporal position in a morpheme). What property of phonemes can be used as a *ground* for such a division?

Evidently, here the *most important property* of phonemes has to be used, this property being perceived *at once,* instantly, because it must be decisive in making a «watershed» between phonemes of the above two kinds. Naturally, the most important property of phonemes is their *being filled by a voice*, i.e., their «*vocality*». It means that all phonemes have to be divided into two classes: *vowels and consonants*, which can be distinguished through their main property: whether a sound is made without audible stopping of phonation — or, on the contrary, a sound is produced by a complete or partial stoppage. These two kinds are well known in phonetics and are designated by simbols *V* and *C* (from Latin words *vocalis* and *consonantis*, respectively). This achievement of the fourth step permits us to go further in the progress of sound structures.

At the *fifth step* (though it is almost impossible to distinguish between this step and the previous one) there exist morphemes as *syllables*, each of them representing a specific «tandem» with fixed positions of vowels and consonants (e.g., the position of a vowel phoneme and the position of a consonant one in a 2-phoneme morpheme, as well as the positions of a vowel and two consonants in a 3-phoneme morpheme, or, vice versa, of two vowels and one consonant).

As it was mentioned by Vitaly Shevoroshkin, 1969; p. 52, «... all the syllabic mechanism has as its aim to provide conditions for the realization of the main element of a syllable — phonation of a vowel. However, a paradox may take place when this main element, this «aim» of a syllable, is absent. Then a voice is replaced by a noise while whispering ⟨...⟩. Nevertheless, the formation of a barrier and its overcoming (sometimes followed by building a new barrier) is the condition that determines the structure of a syllable: the degree of increasing or decreasing of a barrier is the basis for positions of all syllable components before V; and the same scheme is used for the positions of the elements after V; V itself is included in this row as the element with the minimal barrier».

Now we can present examples of such syllables: «*ME*», «*ON*», «*THEN*», etc. Evidently, the *evolution of syllable structures* had to start from 2-phoneme combinations, and only afterwards 3-phoneme combinations (as well as more complex ones) appeared. This conclusion is confirmed by many observations of linguists who mention that a language without combinations «$C + V$» is inconceivable (see, e.g., Trubetskoy, 1960; p. 278). So, what is most natural to humans are *chains* like *CVCVCV*.... Such structures are both the most convenient (this is confirmed by numerous observations on babies' speech) and the initial ones for the development of all languages of the world (Shevoroshkin, 1969; pp. 63–175).

But now it is time (in our construction of the process of evolution, as well as in the evolution itself) to *combine* both *inventions* mentioned above. The result of their «glueing» is, naturally, a sound structure of the *a2b2 type*, responding to the *sixth step* in the evolution. Namely a structure, which has a rather large volume of morpheme dictionary, high reliability, and based on the previous steps of evolution, is destined to be the *final step* of the development of sound structures.

3.7 At the final step of evolution

Nowadays this final step is observed in all natural languages, each of them satisfying all the *requirements* of our construction:

- each language has a sound structure that is based on a set of *differential features*; namely their perception permits the identification of phonemes and morphemes; the *number of* such differential *features* is usually *close to 8*, i.e., responds to the above theoretical predictions;
- *phonemes* («standardized» sound units) are always used by each language; their *number* can change within a very *broad range*, always responding to the predicted range of *14 – 130 units*;
- a rather high degree of *redundancy* (in respect to structures both of phonemes and morphemes) is inherent to each language;
- each language is built of *syllables*, i.e., a more or less fixed sequence of vowels and consonants;
- the *number of* the main *morphemes* in each language is rather *limited* and usually does not exceed *200 – 300 units*.

So, *all* natural *languages* now existing are approximately *equal* in respect to the perfection of their sound structures, though each language chooses its *own principle* of phonemic structure for morphemes, the appropriate amount of phonemes, etc. It seems reasonable to illustrate the results obtained with an example of one of the existing natural languages — contemporary *Russian.*

Several systems describe the phoneme structure of different languages, including Russian (see, e.g., Panov, 1979; Trubetskoy, 1960). We shall base our example on the most widespread description (Panov, 1967; p. 132); according to this, the contemporary Russian language contains 42 phonemes (5 vowels and 37 consonants), which are characterized by 9 differential features. Table 3.2 presents these phonemes and their features, each of them having 2 gradations which are designated by the symbols «+» and «−». (The vowels are considered only in the so-called «strong», i.e., stressed position; in an unstressed position some of the vowels are merged, losing their distinction.)

It is evident that this phoneme structure fully corresponds to the above theoretical conclusions. First of all, it relates to the *amount of differential features.* Though their number is 9 (i.e., slightly more than the predicted 8 units), they are strictly divided into two groups, i.e., vowels and consonants: the feature «consonant» is negative for all vowels and positive for all consonants. So (if we take into account this subdivision, connected with the above syllable structure and related matters) we have only 8 differential features that absolutely coincides with the optimal amount. Besides, the vowels themselves can be described only by 3 features; as for the consonants, they can be described by 7 features (some of them being common with those describing the vowels), but if we exclude the so-called half-vowels м, м', н, н', л, л', р, р', j — then such a feature as «vocality» can be omitted, and we have only 6 features describing 28 «pure» consonants.

The *reliability* of this system is rather high. For instance, the mean value of difference between each pair of vowels is 1.8 features, which provides us with a mere *double distinction*; the appropriate mean value for 28 «pure» consonants is 2.7 features, which provides practically a *triple distinction.* Due to such a redundancy, Russian phonemes are seldom confused with each other. Besides, this is also obtained by a rather clear syllable structure of the Russian language.

As for the *morpheme nomenclature*, the dictionary of morphemes of the Russian language (Kuznetsova and Yefremova, 1986) contains nearly 5,000 units. But among them it is possible to pick out the most productive ones, i.e., those that produce 50 roots or more; we have only *228* of such *productive morphemes*, which is very close to the optimal amount. So, really, the Russian language is «organized in such a way that with the help of a relatively small amount of morphemes (i.e., minimal meaning parts of a word), and due to their constantly evolving combinations, it appears possible to build a great amount of words. And if this property were not inherent to a language, it would not be fit for its task: carriers of such a communicative system (i.e., humans – *V.P.*), consisting of non-recurrent units, would have a memory of a monstrous size» (Kuznetsova/Yefremova, 1986; p. 3).

Phoneme	Vocal	Consonant	Diffused	Low	Flat	Soft	Broken	Sharp	Voiced
а	+	−	−	+	−	−	−	−	+
о	+	−	−	+	+	−	−	−	+
Э	+	−	−	−	−	−	−	−	+
у	+	−	+	+	+	−	−	−	+
и	+	−	+	−	−	−	−	−	+
п	−	+	+	+	−	−	+	−	−
п’	−	+	+	+	−	+	+	−	−
б	−	+	+	+	−	−	+	−	+
б’	−	+	+	+	−	+	+	−	+
ф	−	+	+	+	−	−	−	−	−
ф’	−	+	+	+	−	+	−	−	−
в	−	+	+	+	−	−	−	−	+
в’	−	+	+	+	−	+	−	−	+
м	+	+	+	+	−	−	+	−	+
м’	+	+	+	+	−	+	+	−	+
т	−	+	+	−	−	−	+	−	−
т’	−	+	+	−	−	+	+	−	−
д	−	+	+	−	−	−	+	−	+
д’	−	+	+	−	−	+	+	−	+
с	−	+	+	−	−	−	−	−	−
с’	−	+	+	−	−	+	−	−	−
з	−	+	+	−	−	−	−	−	+
з’	−	+	+	−	−	+	−	−	+
ц	−	+	+	−	−	−	+	+	−
н	+	+	+	−	−	−	+	−	+
н’	+	+	+	−	−	+	+	−	+
л	+	+	+	−	−	−	−	−	+
л’	+	+	+	−	−	+	−	−	+
ш	−	+	−	−	−	−	−	−	−
ш’	−	+	−	−	−	+	−	−	−
ж	−	+	−	−	−	−	−	−	+
ж’	−	+	−	−	−	+	−	−	+
ч’	−	+	−	−	−	+	+	+	−
р	+	+	−	−	−	−	+	+	+
р’	+	+	−	−	−	+	+	+	+
j	+	+	−	−	−	−	−	−	+
к	−	+	−	+	−	−	+	−	−
к’	−	+	−	+	−	+	+	−	−
г	−	+	−	+	−	−	+	−	+
г’	−	+	−	+	−	+	+	−	+
х	−	+	−	+	−	−	−	−	−
х’	−	+	−	+	−	+	−	−	−

Table 3.2 Russian phonemes and their features

Summarizing these data, we can conclude that the sound structure of the contemporary Russian language absolutely corresponds to the most advantageous type (a2b2) — the final point evolution predicted it would reach. A propos, *all natural languages* all over the world arrived at the *same structure*, even though there are apparent colossal differences (including, e.g., Polynesian languages, where vowels make up the overwhelming majority of the phonemes; this is in strong contrast to most of the Indo-European languages, characterized by a rather high concentration of consonants). This is because each language must provide communication with both high reliability and high speed, and therefore must be based on the human operative memory.

3.8 Physical realization of speech communication

The only remaining problem (of the most principal phonetic problems) is to deduce the *physical means* capable of transmitting the above speech communication via the *acoustic channel*. What could be the nature of oscillations of the medium (i.e., the air) that could carry the appropriate sound signals?

Three potential *parameters* claim the role of the main carrier of speech information using sound structures:

1. *Intensity* of a sound signal. This parameter must be rejected for several reasons. First, the value of this parameter (near a receiver's point) depends on the distance between the transmitter and the receiver. Second, it may depend on some individual voice abilities of a concrete transmitting person. And — last but not least — this parameter contains too few gradations (which may be perceived) to carry the rich information corresponding to our structures of type a2b2.

2. The last reason (i.e., limited capacity to convey information) also relates to the *frequency* of the sound signal, if this frequency itself is used as a carrier of information. (It is possible to change this frequency only for the duration of a speech segment, but this relates to the next potential parameter considered below.) Therefore, this parameter must also be rejected, though naturally this parameter — as well as the previous one — is capable of carrying some auxiliary information concerning the spatial conditions of a communication, the individual peculiarities of a transmitting person, etc.

3. *Form* of a signal, meaning its *modulation*, either amplitude or frequency (as well as some combinations of these two versions) for the duration of a speech segment. This parameter appears to meet all the requirements of sound communication; so we shall consider its usage.

Both versions of modulation pre-suppose the existence of a certain «*carrier frequency*», serving as a basis for communication: when using amplitude modulation, the amplitude of these «carrier» oscillations is used to transmit speech information, whereas frequency modulation uses the increase or decrease of this carrier frequency for the same purpose. Let us suppose that we deal with the amplitude modulation (the appropriate analysis applied to the frequency modulation gives approximately the same results.) What can be an estimation for a carrier frequency capable of being used in a speech communication?

This estimation can be based on the following considerations:

- speech information must be received in the regime of functioning of the *operative memory*, i.e., the time for a *morpheme perception* must be *less than .1 sec.*;
- this time must be enough to transmit information concerning merely *8 features* of a morpheme perceived.

We can also suppose (though this approximation is rather rough) that each feature needs approximately equal time to be transmitted, and the transmitting of all the features is successive. [A propos, it was shown experimentally (Panov, 1983; Panov, 1979), that in reality each phoneme consists of several features, which are transmitted in turn; such were the oscillograms relating to different phonemes.]

So, each feature must be transmitted during $.1/8 = .0125$ sec. And during this time interval, a modulation must be superimposed over the carrier frequency in order to express information concerning each given feature. But this requires that the carrier oscillations should occur many times during this interval, e.g., not less than 6–7 times, which responds to a frequency of about *500 cycles per second*. Such must be (on an order of magnitude) the carrier frequency for speech communication.

Now we have to find out whether such frequencies are suitable both to be generated by a human and to be detected.

The *generating abilities* of man seem to be very good for such frequencies. To generate acoustic waves with a frequency of 500 cps, it is necessary to have a resonator of about half a meter in length; these are the dimensions of the human mouth cavity together with the bronchial tubes. And as to the *detecting abilities* — human hearing is absolutely adapted to such frequencies. That is why the human information system was fully ready for sound communication, possessing both the neuropsychological basis for receiving such signals (meaning properties of the operative memory) and the generating-detecting abilities; and when the need for such a communication arose, this communication began its successful development.

* * *

We have considered (using a purely deductive approach) the main mechanisms of language, starting from the necessity of communication and some rather abstract assumptions of an evolutionary character, and, finally, coming to the concrete sounds of human speech.

It was a road from aims and functions — to the mechanisms of their realization, or from «whisper» to «lips» (Osip Mandelstam).

Chapter 4
Construction of the aesthetic sphere

I sit in a wood and stare
Up at untroubled branches
Locked together and staunched as
Though girders of the air:
And think, the first wind rising
Will crack that intricate crown
And let the daylight down.
But there is naught surprising
Can explode the single mind: —
Let figs from thistles fall
Or stars from their pedestal,
This architecture will stand.

C. Day-Lewis, Transitional Poem.

Many various approaches to the problem of links between the spheres of information and aesthetics are possible (see, e.g., Moles, 1958, 1967; Bense, 1969), and one of these approaches seems to be the key to study of culture. This key approach deals with the informational origins, or roots, of different branches of culture. The main problem is to deduce each branch from the deductive model of the human system of information processing, to «construct» the aesthetic sphere by a purely *deductive procedure* (see also Makhmudov and Petrov, 1984; Petrov, 1991/Ref. 7; Sheremetjewa and Petrov, 1991/Ref. 3), as well as to conceive — also by the deductive method — the place of art among the various objects constituting the world of culture.

The starting point of our construction procedure is the trend towards a *more perfect functioning of the human's information system* and an increase in the efficiency of its functioning. Getting ahead of our story, we have to say that at the final step of the forthcoming construction, we shall obtain a number of various means to realize this trend (towards more perfect functioning) such as the phenomenon of women's beauty, objects of design, works of art, etc. So, d'après Paul Verlaine, «De la musique avant toute chose»!

4.1 Natural objects: Absolute and conventional criteria

While constructing, we shall use the so-called method of *binary (dichotomic) classifications*. This method possesses an unquestionable advantage of completeness: it makes it possible to embrace all the cases and situations that are possible in principle (because at each step of the binary classification the whole universe is divided into two parts, and there is nothing outside that); so no class or type of object can possibly be forgotten. That is why this method seems to be the most adequate for the problem studied.

The first basis of our forthcoming classification is quite natural and is connected with the main feature of the objects to be studied, i.e., their *origin*, meaning their *relation to human activity*. The above trend towards more perfect functioning can be satisfied by contacts with objects belonging to two main classes (the classification steps are illustrated by Fig. 4.1).

A. Objects that have *no direct connection with human activity*, i.e., they exist independently of this activity; these objects may be called «natural».

B. Objects that are *directly connected with human activity*, i.e., they exist due to this activity; such objects may be called «artificial».

Of course, there is no fundamental border between these two classes, and there are many reasons for such vagueness. First, it is impossible to determine a «direct connection with human activity» (e.g., sometimes it is not clear whether a landscape is purely «natural» or whether it is a result of cultivation). Besides, the definition of «human activity» is also not so evident (for instance, the human himself can be considered both as a result of human activity — meaning the appropriate actions of his or her parents — and as a result of the activity of Nature). In any case, the border between these two object classes is not strictly determined. Nevertheless, we can use these terms (A, B) because our investigation deals not with objects themselves but with their *perception*. And in this relation, the division of objects (into the two classes mentioned above) is much easier: in every case an object is perceived either as natural or artificial (or maybe sometimes as either «close to natural» or «close to artificial») with the appropriate criteria inherent to the evaluation of the objects of these two classes.

Let us consider the first of these classes (*A*). We have to remember that our main question must be: how can the objects of this supposed class influence their perceivers, and assist them in perfecting the functioning of their information systems?

The answer is almost evident: the appropriate objects have to provoke the more or less perfect functioning of the perceiver's information system by giving it a maximum mutual information between the subject and the object perceived (see Chapters 1 and 2, as well as Golitsyn and Petrov, 1990, 1991). In turn, this requirement supposes a rather high degree of *structural order* in the object perceived. But this order itself depends on properties both of the object perceived and a subject (perceiver). That is why we can stress one of these two sides of perception and, hence, introduce the next (second) ground of our classification.

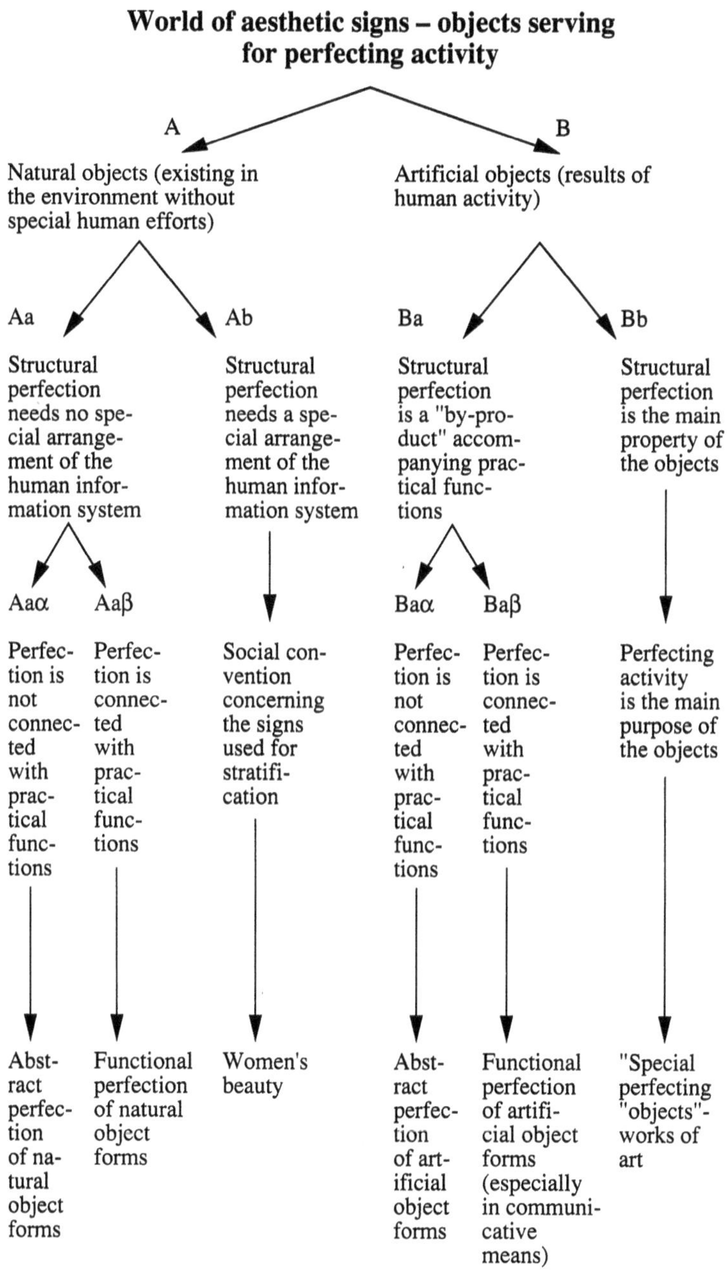

Fig. 4.1 Deductive constructiбn of aesthetic objects

This second principle may be called the *«objectivity»* of the structural ordering of the natural objects perceived. According to this principle, an investigator can divide all objects of the class studied (A) into two types:

Aa. Objects whose structural ordering *needs no special arrangement* (adjusting) *of* the human *information system* to be perceived; in other words, the *criteria* of ordering applicable to such objects are *«absolute»*, natural, non-conventional.

Ab. Objects that *need a special arrangement of* the human *information system* for a human to perceive their structural order; in other words, the criteria of the ordering of such objects are not so evident, «absolute», natural; these *criteria* are rather relative, depend on the special attitudes to such objects and, therefore, may be called *conventional*.

Again we have no unbridgeable gap between these two types (*Aa* and *Ab*) — like in our previous step of classification; but this seems to be not so important for our analysis, because we talk mainly about «polar ideal types» and not about each concrete real object.

So let us dwell at length on the objects of the first type (*Aa*). What might be the means to achieve the appropriate structural ordering of the objects? The nature of this ordering may depend on the *nature of the correlation* within different elements (or properties) of the object perceived; in any case, this correlation has to be rather high.

When speaking of an object we, at the same time, mean a perceiver who observes the correlations mentioned. But a perceiver may consider every object in a different manner, depending on many obstacles, and first of all on the functions of the object observed. So the *functional character* of the correlations may be the next principle in our classification. In accordance with this principle, our objects (*Aa*) are divided into two groups:

$Aa\alpha$. Objects where the correlation observed is *not connected with* any *function* of the object; in such cases we deal with non-functional, *«abstract» ordering* within objects.

$Aa\beta$. Objects where the correlation observed is *connected with* some *functions* of the object; such cases respond to *«functional» ordering* within objects.

Hence, the structure of objects belonging to the first group ($Aa\alpha$) has to be ordered simply in relation to elements of this structure. As far as these objects are directly natural (A) and non-conventional (*Aa*) — one can speak only about the phenomenon of a rather *abstract perfection of forms* of natural objects, e.g., the symmetry of their elements realizing itself, for instance, in flowers, crystals, etc. Another case of such abstract formal perfection is the phenomenon of the so-called «golden section»: if an object possesses a proportion between some of its elements close to 1: .62, — then the form of this object (meaning the above proportion) results in a rather highly positive emotion that has been known for over 2000 years. (For an explanation of the phenomenon of the «golden section» from the informational standpoint see Chapter 5 and Golitsyn and Petrov, 1991.) Many other variants of formal perfection of this kind are possible.

As to the objects belonging to the second group ($Aa\beta$), their structure has to be in good agreement with their functions; this phenomenon can be called the *functional perfection* of object forms. One can find many examples where such functional perfection realizes itself, for instance, in forms of different objects that are very adequate for appropriate function (e.g., biceps of an athlete suitable for his work), in forms of different phenomena that are in ideal accordance with their objective purposes (e.g., a perfect trajectory of a lion's jump to its prey), etc.

Turning to the next type of object (*Ab* — natural objects, conventional criteria), one has to take into account that a *special arrangement* of the human information system evidently may not be applied to all objects, the appropriate type cannot be too broad. Really, in the opposite case (i.e., if this type embraced many kinds of objects) the convention — about a special arrangement in order to perceive the structural ordering of the objects — could come into conflict with other practical needs that require quite different attitudes while perceiving these objects. So, a circle of objects of this type has to be rather restricted, *socially regulated.* What can we find in this circle: flowers or dogs, mountains or people, serpents or birds, clouds, etc.?

When talking about social regulation, one has to take into account that this supposed regulation must fit well into the entire system of social relations, e.g., the links among the people belonging to a society. In other words, our supposed type of object has to be also a carrier of some *socially regulated motives*; these two kinds of objects' influence (on the perceiver) have to accompany each other, being combined in the same objects. But what are these socially regulated motives?

The nature of these motives becomes clear when we take into account that the perception of the proposed objects has to be accompanied by *positive emotional reactions* (arising because of the additional ordering of a human information system). This means, in turn, that some positive estimation must be inherent to the objects of the circle sought for. In other words, within this circle a perceiver can distinguish different objects and estimate them according to an *ordered scale*, e.g., «better — worse», and such estimation has to be rather important — from a social point of view — for the members of the society. Another rather essential obstacle that we have to take into account is that the proposed objects have to be natural. Is it possible to see this group of objects as an area of overlapping of the above two requirements: *to have an ordered scale* important for the members of a society and *to belong to the world of Nature*?

Perhaps several circles of objects can satisfy these two requirements. Nevertheless, one circle is evidently the most suitable for the purposes of our construction, the objects of this circle having the following features:

- in the whole world of natural objects they are the *closest to the perceiver*;
- their properties seem to be very important (from a social point of view) for the members of a society, because they can be used for *social stratification*;
- their properties have to provoke rather strong positive *emotions required by* some *physiological and psychological needs.*

It is not difficult to guess that we mean some *features of human's appearance*: they satisfy all the above requirements. The appropriate objects, i.e., human beings are, of all natural objects, the closest to a perceiver. If the properties of these objects permit us to order them, this ordering (ranking) can be used to distinguish persons occupying higher or lower places in the social hierarchy; the emotion accompanying the perception of such objects — if they belong to a sex opposite to the perceiver's one — can be used for sexual attraction and related purposes.

The last obstacle permits us to narrow the circle of objects to be considered. Evidently, the features provoking the emotions studied have to be *connected with sex*, characterizing the *attractiveness* of a given person for people of the opposite sex. Is there a symmetry between males and females in this relation?

Here we have to remember that our features to be constructed are the product of *social convention*; they are not «absolute» from the point of view of their practical usefulness. Moreover, they cannot conflict with requirements that arise from a practical activity of the society. So, these features are auxiliary in comparison with other, more important properties of humans, and only for some humans are these features rather important. But this conclusion means that the features proposed are not essential for representatives of the «strong» sex (these representatives form a social hierarchy according to quite different properties, mainly connected with practical activity), but rather for *representatives of the «weak» sex* (whose activity and related properties are not so important from the practical point of view). And as far as in the majority of societies, the «strong» sex (dominating in social relations) refers to males — so *females* have to be the carriers of the features studied, and these features may be identified with features of *women's beauty* (see also Golitsyn and Petrov, 1991).

As a result of our deductive construction, this type of object (*Ab*) corresponds with the following situation. It is desirable that in a society among the different hierarchies (characterizing its social structure), one rather specific hierarchy should function that takes into account some of the features of *women's appearance*. There exists a definite *social convention* that indicates *which parameters* are to be considered as valuable («marked», «marqués»), as well as *what meanings* (gradations, values) of these parameters are to be highly appreciated. (For example, the shape of a woman's eye or ankle may be treated as valuable — or not. If this shape is valuable, a perceiver has to know which shape is better, for instance a thin ankle is more valuable than a thick one — naturally, within a definite range of this parameter.) Of course, these criteria, being purely a product of social convention, may have no other objective base but this convention. But every «currency» functions better if it is based on some special properties of the material used. (For instance, gold is used as a universal money partly due to its pleasant colour, glitter, etc.) So these «marked» features of women's appearance may use some *informational criteria*, i.e., the best features may respond to the criterion of the information maximum. (For example, when speaking of a fine figure, the proportion of the golden section is preferable, so the waist has to divide the length of the body).

These criteria are used by the members of a society to determine the value of a woman (mostly of a definite age, usually rather young) and, respectively, the social (hierarchical) position of the man who is connected with this woman (e.g., husband, bridegroom, lover). In turn, men feel delight and other related emotions when looking at such women, and because of this the sexual activity of men acquires positive emotional colouring. In other words, society penetrates humans' souls, and influences their most intimate motives. That is why women's beauty plays a rather important role in society, by combining aesthetic, social, psychological, and even physiological factors on the appropriate informational basis.

4.2 Artificial objects: By-product and special perfecting impacts

But we must not forget the second class of objects (*B*) directly connected with human activity. To classify these «artificial» objects, it seems reasonable to use, first of all, the principle that is of prime importance for all objects specially made for functioning in society — the answer to the question: Is the perfect functioning of a perceiver's information structure the *main worry* of such objects or not? According to this principle, all «artificial» objects can be divided into two types:

Ba. Objects having the above *«perfecting» impact* on the human information system, but this impact is only a *«by-product»*, only an effect «accompanying» some other functional needs in such objects.

Bb. Objects whose «perfecting impact» is their *main function*; naturally, such objects can carry some other loads, but only as «by-products» of the above main function.

Of course, the attribution of each concrete object to one of these types (*Ba* or *Bb*) can depend on a given perceiver — for one subject this given object satisfies only practical needs, and he or she receives no (or almost no) «perfecting impact», while for another subject the same object gives mainly «perfecting» impulses and satisfies no (or almost no) practical needs. Nevertheless, this division is principal for the purposes of our construction, because even this division determines the main structural features of objects to be constructed.

«By-product perfecting impact» inherent to objects of the first type (*Ba*) strictly reminds us of the quite analogous effect of natural objects that need no special arrangement of the human information system (type *Aa*). Here one has to speak again about the structural ordering of the object, correlations within different elements (or properties) of the object, and the *functional character of* these *correlations* that has to be used as the basis for the next step of our classification. Respectively, we are to construct two groups of objects:

Baα. Objects whose *practical function and «perfecting correlation» are not connected* with each other.

Baβ. Objects whose *practical function creates a base for «perfecting correlation»*.

In both cases we deal mostly with the «strengthening» of the practical usefulness of our artificial objects. Again — as was in the cases of groups $Aa\alpha$ and $Aa\beta$ — the appropriate objects have to be ordered either simply in relation to the elements of their structure (group $Ba\alpha$) or in relation to their main functional properties (group $Ba\beta$). The objects with such ordered structures are more pleasantly perceived (they provoke positive emotions due to these ordered structures), and impulses of «structural perfection» come as by-products of the practical functioning of these objects. Examples of objects of the group $Ba\alpha$ belong to the decorative arts: furniture, serviettes, curtains, etc.; in all these cases we observe the abstract perfection of forms. Examples of objects of the group $Ba\beta$ belong to the sphere of design: cars, control panels, bridges, etc.; in these cases we deal with functional perfection of object forms. (Sometimes it is impossible to distinguish between these two groups, because both abstract and functional perfection may be inherent to the same objects.) One should mention that an entire *subgroup* ($Ba\beta 1$) of rather specific objects, where the «strengthening» effect of structural ordering is used by objects destined for *communicative purposes*, includes church preaching, advertisements, newspaper articles, etc.

The most interesting is the last type (Bb), whose objects are destined *especially for «perfecting impacts»*. Evidently these objects can be identified as *works of art*: music, painting, literature, etc. Naturally, such objects can be used also for other purposes; e.g., literature can be used as propaganda of political views, but this property is not «genuine», «inherent» to art, though practically many works of art may function in the society only due to such by-product effects. But, in principle, objects of this type have their *main specific function* which consists in the *«training» of the perceiver's information system*, his/her «emotional training», according to Vygotsky (1971). This function leads to the formation of the world of objects specializing in different kinds of «perfecting impacts», i.e., *different kinds of art*.

When speaking of «training» the person's information system (meaning its more perfect functioning) by some «special objects», we have to consider the main *needs* of this system. So we have to take into account the structure of this system and the processes that take place within it.

The human system of information processing can be considered (Maslov, 1983; Golitsyn and Petrov, 1990) a *multi-level hierarchical structure* (see Figure 4.2). Functioning of this structure consists in the escalation of the information received, from one level to another. Each M-th level possesses its own mechanisms (a set of rules, or paradigm) to process the received information I_M, as well as its own criteria K_M that formulate which information should be transmitted to the next ($M+1$-th), higher level. In turn, this higher ($M+1$-th) level gives (descends) to the preceding, lower level the criteria K_M of choice concerning information necessary for this higher level. Recently, the processes in such a structure were identified with the specialization of two hemispheres of the human brain: the left hemisphere is mainly occupied with the processes within each given level, while the right hemisphere deals mainly with inter-level communications (see, e.g.,

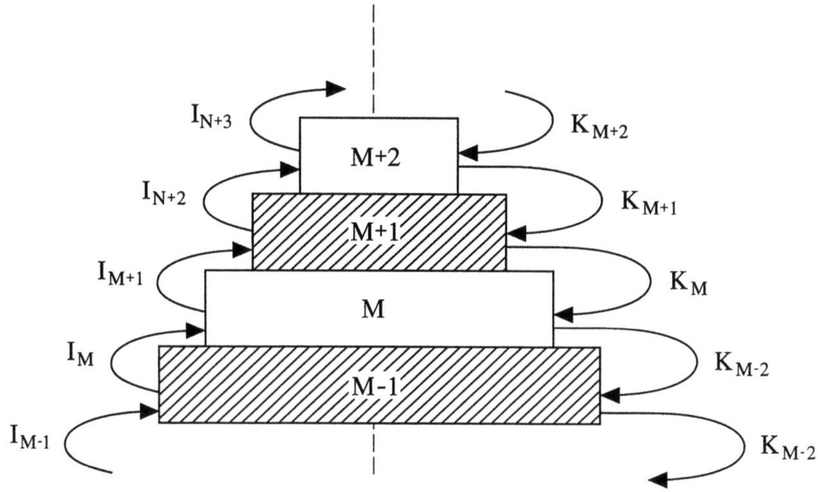

Fig. 4.2 The multi-level hierarchical structure of information processing. Each M-th level processes the information received (I_M) and transmits part of it to the next, $M+1$-th level, according to the criteria of choice (K_M), formulated by the next ($M+1$-th) level.

Maslov, 1983). (Such a structure resembles a multi-level administrative organization where employees of each given department process the information received and transmit the most substantial conclusions to their chief. The chief, working together with the chiefs of other departments — of the same level — processes this chosen information and transmits the most substantial conclusions upwards, to the chief of the next level, etc. While this processing goes on, each employee is working according to his own paradigm and, at the same time, participates in formulating the criteria of processing for his subordinates.)

So, *two regimes* of information processing constitute the nuclear principle of this model: *working within a given level* of the hierarchical structure, and *inter-level transmitting* of information.

Such a hierarchical structure permits us to draw a set of additional conclusions. First, it becomes possible to deduce the main *features of memory devices* that are necessary for the functioning of this hierarchy. (These main features are the usual principal parameters of every memory device, i.e., its operating speed and its volume.) Three types of memory devices respond to the processes inherent to such a structure (see also Golitsyn and Petrov, 1990):

1.) *Memory serving to receive current information.* The main requirement for this type of memory is, of course, its *high speed* of operating. It is especially important in the so-called extreme situations when fast information processing determines even the life of an organism (e.g., when a man is walking in the forest, he has to be able to recognize — with the help of his memory — which beast of prey rushes upon him!). It is necessary to have an opportunity to extract information — which is needed at a given moment — quickly from the memory device, in order to compare this extracted information with the current one. As to the *volume* of this type of memory, the usual regularity may be applied to determine this volume: the bigger this volume, the less the speed of information extraction from it. So to provide a high speed of functioning of this type of memory — its volume has to be *minimal*. It has been shown (using as the base the concept of 2- and 3-gradation coding of information — see Chapter 3, and Golitsyn and Petrov, 1990; pp. 109–118), that this volume is to be $2^3 = 8$ units. And it is known that a human has *three steps* of memory (see, e.g., Atkinson, 1980; Hoffmann, 1982; Klatzki, 1975), and the *first step* is the so-called «*operative memory*» with a rather small volume of about 5–9 units working within time intervals of about .1 sec.

2.) *Memory serving to formulate the criteria of choice* for the lower level. The requirements for this type of memory are quite opposite to those ones in the previous case: these criteria are to be based on the entire experience accumulated by the given level, meaning both the individual experience of a given organism and the experience of his/her predecessors (genetic information). So here the main requirement is the rather *large volume* of the memory device. The speed of the functioning of this device is not very important, because information processing within it takes place all the time, its results being handed down from time to time (in the form of the new criteria of choice) to the lower level. Experimentalists know this type of human memory: it is the so-called long-range memory, or the *third step* of memory. It has practically unlimited volume and keeps information received both recently and very long ago.

3.) *Memory serving to transmit information* to the higher level. The appropriate processes make demands that are intermediate between the ones in the two previous cases. On the one hand, it is desirable to memorize and transmit upwards a rather large amount of information, though not too large (in order not to overload it by superfluous information). But, on the other hand, transmitting of this information is not to be delayed, because otherwise all the activity of the information structure (as well as the activity of the appropriate human executive mechanisms) would be very slow, and this may be dangerous, especially in extreme situations. That is why to provide the processes of this type, it is desirable to have a memory device with *intermediate parameters* (concerning both the volume of keeping and the speed of operating) between the two previous types. And experimentalists are familiar with this type of memory: it is the so-called *second step* of memory; its volume is much more than that of the first step («operative memory»), but it works more inertially, within time intervals from one-tenth of a second to several seconds.

So, the basis of our construction consists of a multi-level hierarchical structure with two types of information processes and three types of memory devices. What could be the means to «train» such a system, to make its functioning more perfect?

We suppose that the means (used by the appropriate «special objects») to be found are to be rather different for the different types of the above processes.

Returning to the «*perfecting needs*» of our hierarchical structure, we have to distinguish two types of such needs (and each type we hope to connect with the appropriate type of perfecting «special objects» to be constructed):

$Bb\alpha$) needs to deal with the *processes of a given level* of information processing;

$Bb\beta$) needs to deal with *inter-level communications*, transitions between different levels of the structure.

But to clear up the mechanisms of satisfying these needs we have to consider the «*terms*» that are used by this system while processing information. These «terms» are functioning at each level of the hierarchy though, of course, they may be different, changing when going from one level to another. And it appears (see also Chapter 3, and Golitsyn and Petrov, 1990; pp. 100–104) that at every level *two sets of «terms»* should be advantageous both to store the information and to process it:

— A set of *concrete images* corresponding to real objects of the surroundings that are important for a given person. This set of images also should be multi-level, e.g., at one level such images may be John, Peter, Mary, Lilian, etc.; at another level such images may be a man, a woman, a boy, a girl, a baby, a child, etc.; and we can find a higher level where such images are: a human, a tree, a dog, a wolf, etc.

— A set of rather *abstract features* relating to some properties that are common for different objects of the surroundings, these properties being important for a given person. This set also should be multi-level, e.g., at one level such features may be a colour (white, red, green, etc.), dimensions of objects (their length and width), loudness, etc.; at another level such features may be: fitness for food, attribution to the organic or inorganic world, to the world of flora or fauna, etc.; and we can find a level where such features are a degree of a human consciousness, humanistic attitudes, etc.

These two sets of «terms» are very essential for our «special objects» to be constructed, because the *characteristics of* our *objects* are the main purpose of the forthcoming consideration. And, evidently, these two sets of «terms» have to generate two main variants of means capable of making the functioning of our hierarchical structure more perfect. (Both variants may be inherent to any level of our hierarchical structure.) These variants serve for:

$Bb\alpha_{\text{im}}$) training operating with concrete *images* of objects;

$Bb\alpha_{\text{fe}}$) training operating with abstract *features* of objects.

Now we have at our disposal all the «raw materials» to construct the system of arts. We are simply to *combine* the appropriate *conditions* (functioning within a given level or, vice versa, inter-level transitions) *with* the appropriate *mechanisms of memory* (its three steps) *and* the appropriate *«terms»* (concrete images or abstract features) with each other. As a result of such combining, we receive different «alloys» of perfecting that hopefully will correspond to different *kinds of art*: music, fine arts, prose, verse, etc. In principle, the number of possible combinations is equal to $2 \times 3 \times 2 = 12$, to say nothing of the colossal number of potentially possible combinations of the modalities of the above factors and ways of their practical realization — so the resulting amount of combinations should be simply gigantic! But we may suppose that only some of them can exist in real practice, because the rest of the combinations should be not «self-consistent». So to find out *which* of these possible *combinations* are *self-consistent* and what the *characteristics of* the appropriate *objects* should be (works of art, belonging to different kinds of art) is the next step of our deductive construction.

What should the optimal strategy of such a deductive construction in the case of the system of art be? Evidently, the best strategy is to take into account different obstacles (factors) in a sequence that responds to their significance, i.e., to consider first of all the most important obstacles, then slightly less important ones, etc. Such a sequence (something like the introduction of a series of «disturbances» in theoretical physics) can provide, first of all, the principal characteristics of «special objects» to be constructed and then their different versions, details, etc. Such a procedure of the deductive construction was accepted to receive the characteristics of «special objects» responding to different kinds of art.

4.3 Concrete images and art

It seems reasonable to start from the main division of all the processes that are inherent to the human multi-level hierarchical structure of information processing (Maslov, 1983; Golitsyn and Petrov, 1990). As it has been shown, this main division deals with the functional localization of informational processes that can take place either within a given level of the hierarchical structure ($Bb\alpha$) or between neighbouring levels of this structure ($Bb\beta$). Evidently, the first of these two types of processes is the «basic» one, because without it further transmitting of information is senseless. That is why we are to begin our construction from the first ($Bb\alpha$) type of process, i.e., from the *processes* taking place *within a given level.* (The logic of the considerations below is illustrated by Figure 4.3.)

According to the logic of deductive construction, we are to combine the above division with the next division, i.e., the division of the «terms» that work at each given level: concrete images or, vice versa, abstract features. Even this division should now be used, because it concerns the most important aspect of the forthcoming consideration, i.e., the nature of the features of «special objects» to be constructed.

When speaking of operating with *concrete images* ($Bb\alpha_{\text{im}}$), we have to take into account that in our everyday life the use of such images is nothing else but *orientation in* the real *sensuous space* where the person lives, i.e., in his/her surroundings — in surroundings consisting of different real objects. So in the frame of this variant of the «perfecting activity» two ways are, in principle, possible:

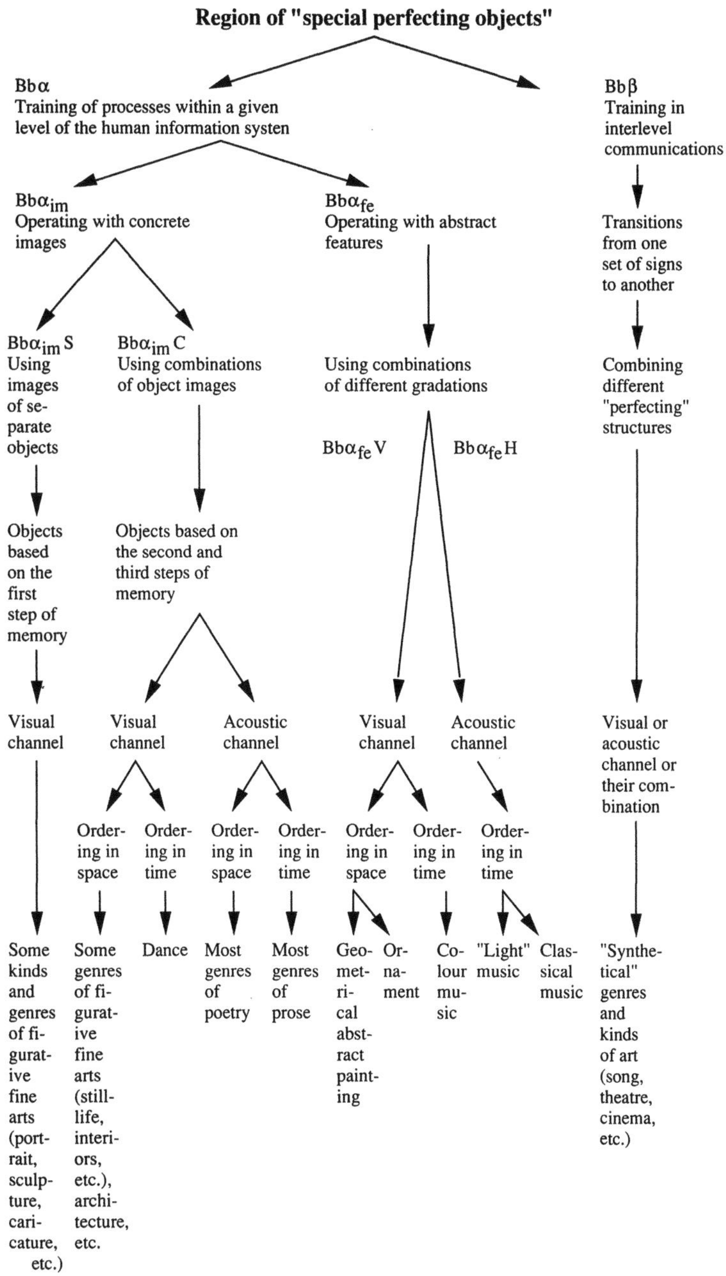

Fig. 4.3 Deductive construction of «special perfecting objects».

$Bb\alpha_{im}S$) to use images of *separate objects*;

$Bb\alpha_{im}C$) to use *combinations* of object images.

(Each of these two ways includes both a *reproduction* of images of different objects familiar to a given person, e.g., images of a given dog, a dog in general, an oak, a tree, a jump of a lion, etc. depending on the level considered — and a *creation* of images that are new for a given person. But no fundamental border exists between the reproduction of images and the creation of new ones: on the one hand, every familiar image can differ from the appropriate familiar object, i.e., the image's prototype, more or less slightly; on the other hand, every new image is created — and, much more important, perceived — on the base of the familiar images; the border between familiar images and new ones — or the border between images that may be called or interpreted as old or new — is rather flexible, its position depends on the recipient's psychological attitudes. That is why we shall speak further of images reproducing different objects or their combinations, while meaning also that they create new images or new combinations of images. Besides, the absence of a border between reproduction and creation may appear to be rather useful for the evolution of the sphere of the appropriate «special objects».)

What could be the practical possibilities of using these two ways as instruments of our «perfecting activity»? The principle of the information maximum permits us to answer this question: the recipient's information system is maximally affected when the new object perceived differs from the appropriate familiar object (its prototype), and the *difference* between these two images is neither too small nor too large; it should be *optimal*. (This optimal value of their difference can be calculated on the basis of the principle of the information maximum — see Chapter 5.) Due to such difference, a person's system of information processing receives an additional impulse for activity, and a rather specific «training» of this system takes place. We shall try to apply this result to our possible ways of operating with concrete images.

So taking into account the possibilities of realizing in practice the first way to deal with *separate* object *images* ($Bb\alpha_{im}S$), we can construct the following *devices*:

— The appropriate *choice* of objects to be reproduced by their images. In this case we can use only the properties of the statistical distributions, both of objects and their images; in other words, statistical distribution of the objects chosen should differ from the statistical distribution of the objects the person has met in his/her experience. For instance, if leopards are seldom met in the given locality, it is desirable to reproduce leopard images more or less frequently in order to receive the optimal difference between statistical distributions of such objects and their images;

— The appropriate, rather well pronounced, *distortions* of images in relation to their object prototypes. In this case we use the structural difference between a real object and its image; for example, some parts (details) of the object can be enlarged, or an image can reproduce an object without some of its parts (details), etc. This permits the receipt of the optimal difference causing maximal influence upon the recipient's information structure.

The second way to deal with *combinations* of object images ($Bb\alpha_{\text{im}}C$) can be realized in practice only through either the previous devices or through uniting the chosen images into appropriate combinations which differ optimally from the object combinations familiar for a given recipient. For instance, a leopard and a bookshelf are never met together, and this combination would be too sudden for every recipient; and vice versa, a bookshelf and books are usually met together, and this combination is too «trivial» for every recipient. So it is desirable to find a combination of «something» and a bookshelf in order to receive the optimal difference between the combination represented and the appropriate combinations a person has met in his/her experience.

Now we shall step aside the «mainstream» line of our discussion, in order to answer the question: what may be the *interrelations* between the above ways ($Bb\alpha_{\text{im}}S$ and $Bb\alpha_{\text{im}}C$) of operating with concrete images?

First, evidently *no* fundamental *border* exists between images and their combinations, the appropriate distinction depending mainly on the person's psychological attitudes. (Quite the same situation took place just above, in the case of the distinction between the reproduction of images and the creation of new ones.) For example, a table, a chair and a bookshelf placed in a room can be perceived, by a given person, separately, as different objects — if the person is prepared simply to identify them; and vice versa, the same objects in the same room may be perceived together, in their integrity, as an image — if the person treats them as a variant of the space organization of a scientist's cabinet, or as a gestalt image of the entire intellectual sphere. So, there are no fixed (hard) borders between these two ways — transitions between them are possible — that would appear rather substantial for the *evolution* of the sphere of the appropriate «special objects».

Second, both ways offer definite *limitations* to the objects that can serve as «prototypes» for the images to be used by the appropriate «special objects». These limitations are caused by the *mechanism of comparison* which needs to receive a maximum of information due to the optimal (or close to optimal) difference between an image and its prototype. Thus, simply reproduced images, if only the appropriate choice of objects is used, suppose that their prototypes are not too widespread in the real experience of the recipients. And such limitations appear especially important when the entire *combination* of images represented should definitely differ from the appropriate «prototypical combinations». Here we have nothing else but the requirement to deal with quite definite (discrete) *«standardized» object combinations* that are rather frequently met in the experience of a recipient (in the opposite case, the above comparison becomes impossible). This conclusion will be rather substantial for further construction of the concrete *genres* within different kinds of art.

But it is time to return to the main line of our deductive construction.

Now we are to draw on the remainder of our «raw materials», i.e., the *steps of memory*, in order to make a jump from different kinds of operating with concrete images to the properties of the appropriate groups of «special objects» — these groups to be further named as kinds of art or their genres. To solve this problem, it seems reasonable to find out which steps of memory are mostly connected — in the practice of the person's orientation in his/her sensuous space — with the ways deduced above. And considering the steps of memory, we have to take into account their principal characteristics mentioned above, i.e., the operating speed and volume of the memory device.

The first way dealing with more or less *separate, elementary objects* ($Bb\alpha_{im}S$) usually requires — in real practice — mainly a *fast-reaction*, in recognizing a perceived object, etc. It is the main requirement here, and against its background all other considerations seem to be much less important. So for the appropriate «special objects», the *first step of memory* should be the most suitable.

In turn, the fact that the first step of memory is the probable basis for the functioning of a group of our «perfecting special objects» permits us to deduce some conclusions, the most principle one concerning the *information channels* which might be used by such «special objects». In fact, a person possesses several information channels, e.g., hearing, vision, sense of smell, sense of touch, feel of gravitation. Which of them can be used for the objects to be constructed?

As far as we have earlier used such a characteristic as high speed of reaction, we should use it again to find the suitable information channels. But only one channel, namely *vision*, can provide the fastest receipt of information. That is why the given kind of «special objects» should use mainly the visual information channel; all other channels are much more inertial and cannot be compared with vision.

What should be the *structure* of such visual «special objects»? They are to reproduce visual concrete images of objects belonging to a recipient's sensuous space. They are also to be based on the first step of memory, so such «special objects» should be *immobile representations* of some of the concrete objects of this space. Such objects may be identified with the works of some genres of the so-called *figurative fine arts*, e.g., some genres of figurative painting and sculpture. To concretize the structural features of such «special objects» (i.e., to receive some concrete genres of fine arts) we are to concretize further the factors that we have used while building our construction. The factors we should concretize are the following:

- the *devices* used while realizing in practice the given way of operating with separate images;
- the possible *modalities* of the used information channels, i.e., vision.

The devices that can be used by the given way (i.e., the appropriate choice of objects and the appropriate distortions of images) were considered above. As to the possible modalities of vision, the last possesses several ones destined for person's orientation in the sensuous space, e.g.:

- modality serving for *black and white vision*, in the conditions of weak illumination (almost in the darkness); this modality exists due to special non-spectral light detectors in the human eye;
- modality serving for *colour vision*, in the conditions of good illumination; this modality exists due to three types of spectral light detectors in the human eye;
- modality serving to determine *short-range distances* to the object perceived, in the conditions when this distance is less than 1.5–2.0 meters; this modality works due to accommodation inherent in the lenses of the human eye;
- modality serving to determine *long-range distances* to the object perceived, in the conditions when this distance is much more than 1.5–2.0 meters; this modality works due to convergence of the eye's axes.

Combining these concrete modalities with the above concrete devices, we can obtain different *groups of «special objects»* which should respond to some *concrete genres* (in the broad sense of this word) of fine arts. Thus, a combination of the appropriate choice (of an object to be reproduced) with the modality serving for colour vision and the modality serving to determine the short-range distances (to the object perceived), gives such a *genre of painting* as *portraiture*. (Here we omit the intermediate considerations leading to this result, i.e., taking into account obstacles that accompany the use of the above devices and modalities, e.g., the fact that to use a familiar object may mean to reproduce the image of a person because a person is usually observed at short-range distances, etc. We can find another variant of the realization of the same combination of devices and modalities satisfying some other accompanying obstacles in such a «genre» — in the broad sense of this word — as *sculpture*.) Some other genres of figurative fine arts can be deduced with the help of such analysis, e.g., *caricature* (which responds to rather pronounced distortions of objects combined with the modality of black and white vision).

In the framework of the second way dealing with *combinations* of object images ($Bb\alpha_{\mathrm{im}}C$) we have to consider the situations responding to:

- uniting images according to their *location in the space*;
- uniting images according to their *sequence (ordering) in time*.

(Of course, both kinds of unification can exist together, but these two «polar» kinds are the «nuclear» ones in all the cases, and hence should be considered first.)

Because this way concerns a certain amount of objects — here the main requirement is a certain *volume* of memory — this volume can be provided only by the *second step* of memory or the *third* one. The limitations connected with the operating speed are not so important here, and they do not determine the information channels to be chosen for the appropriate «special objects». However, the choice of the information channels now should be determined by their possibilities to give a large «richness», a diversity of object properties. But, evidently, these properties can be ordered, because they should form a definite structure having the optimal difference in comparison with the appropriate «prototypical» structure.

Both such diversity and ordering possibility can be provided by means of only two information channels: *vision* and *hearing*. (True, the sense of smell or sense of touch can provide rather rich, diverse information; nevertheless, this information can be hardly ordered.)

To determine a *structure* of the appropriate «special objects», we should first combine the above situations of uniting images (due to their location either in space or in time) with these two information channels (vision and hearing). And, at once, we receive four kinds of «special objects» responding to:

— Image *combinations* ordered *in space* and *perceived by vision*. Here we can find some genres of the figurative painting, such as *still-life*, *interiors* (using mostly the second step of memory and the short-range modality of vision), as well as *architecture* (using both second and third steps of memory together with the long-range modality of vision);

— Image *combinations* ordered *in space* and *perceived by hearing*. Here the most genres of *poetry* can be found (because the process of reading is nothing else but listening to speech, the signals of which can designate different objects and generate appropriate images), using mainly the second step of memory. Many poetical works give us the sensuous picture of image combinations (spacial situations) during a rather long process of perception;

— Image *combinations* ordered *in time* and *perceived by vision*. One should use dynamic, *mobile representation* of some concrete objects; here the art of *dance* can be found (using both the second and the third step of memory);

— Image *combinations* ordered *in time* and *perceived by hearing*. Most genres of *prose*, which use mainly the third step of memory, are found within this category.

To deduce the main features of each of the above genres (as well as of some other genres), it is necessary to take into account the *modalities* of the information channels used. This is especially important where *hearing* is concerned, and the mechanisms of *speech perception* are involved. As it was shown in Chapter 3, the mechanisms of processing speech information contain *three modalities*:

- the *«associative» modality* to *establish* associative *connections between signs and their denotates and/or concepts*, i.e., between the given word and the appropriate image generated by this word;
- the *«correlational» modality* to *check and establish statistical correlations within* the world of *signs*, i.e., between words in different pairs;
- the *«grammatical» modality* to help the previous modality to function by *indicating the pairs of signs* — words that have *to be checked* as to whether they were met together in the previous experience of a given recipient.

These three modalities usually function together during speech perception, and it seems reasonable to «train» each of them by means of «special objects» (belonging to the prose and verse that we constructed earlier). Naturally, the main purpose of such «training» must be the *joint functioning* of these modalities. Nevertheless, it seems possible to *stress* the functioning of one (or two) of them, and this then results in different concrete *genres* or *directions* within concrete genres. For example, poetical works usually stress the «associative» modality, giving a picturesque view of the environment or of the author's internal world, while prose is mainly based on stressing the «correlational» modality forming rather large networks containing many personages, landscapes, situations, etc.

Further, even within prose it is possible to *use the associative modality* rather *intensively or, the other way round,* to *neglect this modality*, and then *two* quite opposite *directions in prose* emerge. The first direction is aimed at creating a very *sensuous picture* of a real or imaginary situation in the environment, while the second direction is mainly concerned with «*pure symbolic games*», «games of signs» which are almost «empty» in relation to their sensuous component. (In Russian literature, the first tradition develops from Nikolai Gogol's prose, the second one — from Alexander Pushkin's prosaic works. See, for example, the appropriate observations by Eichenbaum, 1959, as well as the results of the measurements of prose «transparency» made by Sheremetjewa and Petrov, 1991/Ref. 4. In general, some arguments support the theory that the first, «sensuous» direction is oriented mainly towards the activity of the right hemisphere of the human brain, and the second, «symbolic» direction is more connected with the activity of the left hemisphere.)

In this way we have come to the *construction of* different *directions* within prose, as well as within the entire region of literature. These directions do not coincide with the traditional genre divisions, but respond to the real processes that take place in the human information system. In principle, it is possible to concretize the details of each of the above mentioned genres and branches (directions) of «special objects» belonging to this variant, but such detailed deduction is not in the frame of this publication.

We have studied all the possible (in principle) opportunities of artificial objects destined to «train» the information processes that take place within a given level of the human information structure and operate with concrete images (BbA_{im}). Now it is time to consider the «training» opportunities that deal with the analogous processes that do not involve concrete images, but their rather abstract *features* ($Bb\alpha_{\mathrm{fe}}$).

4.4 Objects' features, signs, and art

To work with features has become a habit for each person, because this way of thinking is very effective (as it was shown in Chapter 3). That is why (though it is not the only reason) operating with features became widespread in the human information system. Quite different features can be used at different levels of the hierarchical structure: colour of objects, their sizes, their fitness for food, sex of a person, age, his/her intellectual ability, etc.

In as far as *operating with features* is inherent to the human information system, it seems desirable to «train» this kind of operating with the help of some artificial «special objects». And because this task is *common* for different features and their gradations, it seems desirable to train a way of operating with some *signs* (symbols, not images) of a rather abstract character, i.e., to organize a «game of signs», these signs being the abstract features and their gradations. Which are the «special objects» that realize such training?

Here we have to repeat the previous statements we made when we dealt with concrete images. The main difference is that it is *senseless* to use *the only* (one) *gradation* of one feature as a basis for a «special object»: a gradation can be perceived only under the condition that at least two gradations are present within a given object. So each «special object» to be constructed has to be based on a *combination of* different *gradations* of a given feature or on *combinations of gradations* of different features. And these combinations have to be «perfect» in the sense mentioned earlier, i.e., they have to be *ordered* in a definite way to exert optimal influence on a recipient's information system.

This ordering (relating to the signs used, i.e., gradations of a feature or of some features) can take place *in space* or *in time*. (And, again, these two orderings can exist together, but each of them is a «nuclear» one and should be considered separately.)

But both types of ordering can be based on some *properties* that permit (in principle) us to *identify* the *signs* — features and their gradations which constitute the structure that influences the human information system. And, again, only two information channels — *vision and hearing* — possess the appropriate abilities. So, first of all, we have to analyze the problem of the existence of *sign systems* that allow us to create *ordering either in space or in time*, and use *either a visual or an acoustic information channel.*

Earlier, when we dealt with the analogous relationship that did not deal with abstract signs but with concrete images, the problem did not exist. When working with concrete images it is possible to use the visual channel for creating an ordering both in space and time: an ordering in space can be (evidently) created quite simply because of the visual properties of the images used (e.g., the spatial ordering of visual images in figurative painting), while an ordering in time can be achieved because of the duration of the process of visual perception (e.g., the temporal ordering of visual images in the art of dance). Quite similarly, by means of an acoustic channel it is possible to create an ordering of images both in time and space: an ordering in time can be created because of the simple sequence of acoustic signals (e.g., the temporal ordering of images in prose), while an ordering in space can be achieved because of the associations connecting the acoustic signals with the spatial properties of objects (e.g., the spatial ordering of images in poetical works). So, the situation of using visual and acoustic information channels to create either spatial or temporal ordering appears to be quite symmetrical.

Now, when we deal with a «game of pure signs», the situation is slightly different. The difference lies in the fact that in this «game», we cannot use the associations that connect the signs used with some images (denotations or concepts); we can use the *signs themselves*. That is why only *three kinds* of «special objects» operating with «pure signs» can exist: *combinations of visual signs* creating an *ordering either in space or in time*, and *combinations of acoustic signs* creating an *ordering in time*. As to the steps of memory needed, they depend on the procedure of the *identification of signs*, i.e., features and gradations that are used in the «special objects» to be constructed; that is why the appropriate memory devices have to be considered separately for each of these three kinds. As a result, we receive the following groups of «special objects» (see also Petrov, 1993):

$Bb\alpha_{\mathrm{fe}}V_{sp}$) *Combinations of signs* (gradations of a feature or of some features) *perceived by vision* and ordered *in space*. What steps of memory can be used by such objects? Evidently, two versions are possible, corresponding with the appropriate groups of objects:

— Combinations of visual signs supporting the condition that the *identification* of features and their gradations (constituting the structure perfecting the recipient's information system) *needs no* — or almost no — special *time*, effort, etc. In this case, our «special objects» to be constructed can use the *first step of memory*. In other words, such objects have to be rather *simple* — or at least the task of identifying their features and their gradations, i.e., the signs to be used, has to be simple. These «momental» objects, consisting of «pure» simple visual signs, come up (in their «genuine» form) in such branch of fine arts as *geometrical abstract painting*.

— Combinations of visual signs possessing the specific property that the *identification* of features and their gradations (constituting the appropriate structure) *needs* definite *special time*, effort, etc. Here either the *second* or the *third step of memory* must be used. The «genuine» field of such objects is an *ornament* that can be read as a «text» (inscription), and its repeated elements are identified thanks to their distribution in space, requiring definite time for «decoding».

$Bb\alpha_{\mathrm{fe}}V_{ti}$) *Combinations of signs perceived by vision* and ordered *in time*. Here we use a sequence of signs requiring a definite time for perception; that is why it is impossible to use the first step of memory. Therefore, it is not so important if time for identification (of features and their gradations) is needed or not: in any case, only the *second* or the *third step of memory* has to be used. Such dynamical sequences of visual signs are realized in the so-called «*colour music*», i.e., a specific kind of art consisting of «pure» colour elements whose projections on the screen change according to a definite program.

$Bb\alpha_{\mathrm{fe}}H$) *Combinations of signs perceived by hearing* and ordered *in time*. Here again two versions of realization are possible:

— Combinations of acoustic signs supporting the condition that *identification* of the features and their gradations (constituting the structure optimally influencing the recipient's information system) *needs no* — or almost no — special *time*, effort, etc. The *first step of memory* fits the requirements of such combinations. Each of the appropriate objects is nothing else but a combination of acoustic signs possessing a definite order. It is clear that one can find the «pure» realization of such objects in *musical accords* and other *musical phenomena of a short-range scale*; these phenomena are a basis for some genres and directions of «light» music — like *jazz*.

— Combinations of acoustic signs possessing such a specific property that *identification* of the features and their gradations *needs* definite *special time*, effort, etc. The *second* or the *third step of memory* has to be used here. It allows us to use acoustic signs to which simultaneously many features are inherent, and while perceiving the sequence of these signs a recipient can reveal (identify) different features of these signs, so the resulting structure can be «multidimensional» (its elements belong to several structures). This means there is a colossal diversity of possible many-featured acoustic structures, this diversity realizing itself in the many genres and directions of *classical music* (the last is oriented mainly towards various long-range orders, which differ from most genres of the above mentioned «light» music, including jazz). And a special system of abstract signs, i.e., musical sounds, was created for these «games»; the elements of this system have nothing in common with any concrete sounds met in nature (for more information about this property of musical sounds, see also Levi-Strauss, 1964).

These are the opportunities to work with «pure signs», i.e., abstract features of the environment, to make the human information system function more perfectly. Of course, some of the above-mentioned opportunities can be *combined* in the same «special object» (work of art), giving simultaneously several kinds of informational «training». Nevertheless, these *combination possibilities* are rather *restricted*, because during perception a recipient is not capable of «detecting» many «perfecting» structures (within a given «special object»), but only one or two (or three). This fact is the result of our «predilection» to search for quite definite structures. This attitude is like the «focusing», the «tuning» of every information system; in physiology and psychology this phenomenon is connected with the theory of the «dominant». In the framework of our informational approach, at a given moment of perception a recipient is able to keep in his/her memory no more than 8 elementary units of any structure; the simplest structure has to contain at least 2 – 3 distinctive marks, or signs; so the maximal amount of structures perceived simultaneously cannot exceed 2 or 3. That is why we often meet the above groups of «special objects» in their «pure» form (e.g., definite kinds of art, or definite genres or directions are fully devoted to one of the above kinds of «training»), though sometimes the opportunities mentioned are «mixed» within the same kind of art or its genre.

Finally, we have to consider the last type of process in the human information system, i.e., *interlevel communications*, taking care of transmitting information from one level to another ($Bb\beta$). What can be the means to perfect these processes?

Interlevel communications consist in the escalation of information, i.e., *transition from one set of signs* inherent to a given level of the human information system *to another set* inherent to the next, higher level. In other words, here the «heart of the matter» is «*translating*» from one language to another. The best means to train such «translating» is to *combine both languages* in the same object

— in the same way that in teaching a foreign language it is useful to show the student an object and to give him/her the name of this object both in his/her native language and in the foreign one. In the case of our «perfecting special objects», this type of training means nothing but the «hybridization», the *synthesis of* perfecting *structures belonging to different levels* of the human information system. Such a combination of different structures (their «being glued together») can make interlevel transitions more flexible, versatile, etc.; in other words — more perfect.

This way of training leaves us with another question concerning the amount of structures to be combined («glued together») in the same «special object». As was mentioned above, because of the recipient's desire to look for quite definite structures, the amount of these structures has to be rather *restricted*, e.g., to *two or three structures.*

As to the nature of these structures that are «glued together», their nomenclature can be arbitrary, i.e., it seems useful to combine two of them (or three, though in this case the process of perception is expected to be much more complicated). So, by combining different kinds of perfecting structures mentioned above, we can construct various *synthetical* (in the sense of combining different perfecting opportunities) *kinds of art* and their genres. For instance, hybridizing an image combination ordered in space and perceived by hearing (an opportunity which is «purely» realized in some poetical works), and a combination of acoustic signs perceived by hearing and ordered in time («purely» realizing itself, e.g., in «light» music), we receive such a synthetic genre as the *song*. Some effects of a «synergetic» character arise out of such a combination, increasing the total, resulting influence both of the poetical and the musical components of a song.

In general, some combinatorical effects are inherent to almost any kind of art or genre, and moreover, to most works of art. But, in many cases, these combinations are nothing but the simple sum of devices which influence a recipient's information system. And only in «synthetical» genres and kinds of art the «glueing together» of different sorts of training is principally important, generating a new quality of «hybrid special objects».

Many other synthetical genres and kinds of art, which obviously include *theatre* and *cinema*, can be constructed with the help of this combinatorical method. But a detailed construction of the various kinds of art and their genres and directions, as well as of the devices used by them, is the subject of a separate investigation.

The above analysis has permitted us to construct — via a purely deductive method — the foundations of various aesthetic sign objects that help a human to survive in the environment. Besides being of *theoretical* (gnoseological) *interest*, the results received make it possible (thanks to the classification derived) to *measure* the influences of different aesthetic objects, meaning *both* their *pragmatic value* (connected with their usual, practical functions) *and* their specific *«perfecting» role* in the interaction between a human and his/her surroundings.

4.5 Emotional impact of art: Social aspect

By being perceived, works of art exert many influences upon different facets of human' spiritual life. But from the standpoint of system research, it seems reasonable to view these influences in terms of a broader system that includes art as one of its subsystems, i.e., with the social system. So our task is to study the links connecting art with social life as a whole.

The problem of the above links may be approached from two «polar» points of view. The first approach deals mainly with the influence of the social system on the character of art (see, e.g., Martindale, 1990). In spite of the feedback, the social system as a whole (embracing many branches, art being only one of them) has a definite influence on the character of each embraced branch, e.g., synchronizing impulses to «switch» between «analytical» and «synthetical» styles in such different branches as socio-political «climate», economics, art, etc. (Petrov and Yevin, 1991). All these branches have a common basis, i.e., the *emotional sphere*; that is why this sphere plays the role of a specific *medium* transmitting interactions of different branches. Due to this, the second approach seems to be promising, as it deals mainly with the contribution of art to the emotional sphere and, therefore, to social life as a whole. This contribution is now the main topic of our consideration.

The impact of art on social life — what is its *character* and *quantitative estimation*? These two questions need separate analysis.

1. The *character* of the influence of art on social life may be very diverse. For instance, art may play a leading role in relation to some particular fields of human activity. For example, painting may serve as a peculiar «tuning fork» for the perception of colours, or orientation in the «world of colours»; dancing may give criteria for the «world of plastic movements»; poetry creates a paradigm for our «speech world», etc. (see, e.g., Jakobson, 1987). Nevertheless, there must be some more *common influences* concerning the main features of our social life. Our further consideration will be devoted to such «common influences».

To determine their character, we have to start from the point of view of «self-consistency» of the social system as a whole. (In turn, the principle of self-consistency itself may be deduced from the principle of the information maximum: it is precisely that principle that explains the tendency towards the growth of a degree of ordering within a system, this system being either a human or society.) The special emotional medium (substratum), being an object of different impulses, evidently needs unity and self-consistency, and art is one of the means to achieve this aim. So, the main feature of art is to *reconcile contradictions* within the system of consciousness, i.e., to make it self-consistent.

As to the human, this main feature results in contributions to the «harmony in a human's soul» due to the contacts with art. But when speaking about society, which is quite analogous, «*contribution to harmony*» means contributions to:

a) self-consistency of the system of different kinds of art;

b) self-consistency of the social system as a whole.

Naturally, these two contributions are not strictly separated from one another. (Moreover, self-consistency of the social system means a definite self-consistency within the sphere of art, i.e., of different kinds of art.) Nevertheless, these two aspects of harmony are not tautological because any self-consistency is never absolute, and in every case (*a*, *b*) we may stress quite different components of self-consistency.

Self-consistency of the *system of art* (*a*) means that all kinds of art «speak of the same matter» (Woelflin, 1913; p. 78), the only difference caused by the languages used. That is why different kinds of art — as well as different branches of social life — have to show synchronous evolution, and this was observed in music, painting, architecture, the socio-political «climate» and economics (Petrov and Yevin, 1991). Thus, each kind of art gives its contribution to this synchronous process, using the emotional sphere for this purpose. But this phenomenon and the appropriate contributions are not the main topic of the present consideration.

Let us consider the self-consistency of the *social system* as a whole (*b*) as the purpose achieved, with the help of contributions of different branches of social life. Evidently, here, first of all, one has to take into account the *main social ideas* dominating in society. The main features of art cannot be in conflict with these main social ideas. (We mean the majority of contributions resulting from art perception by the majority of the members of the society. Naturally, some contributions of quite different types have to take place, these contributions being a result of the necessary diversity within each kind of art. In turn, this diversity is necessary both for the current functioning of art and for its development, as a reserve for future changes of art.) This *accordance between art and ideas dominating in the society* may concern different levels of consciousness. Evidently, there is no need to duplicate strictly concrete social ideas by means of art, but at some rather high levels of generalization (maybe far enough from the level of concrete objects and phenomena) the appropriate accordance has to take place. For example, if in general the spirit (the emotional colouring) of social life is optimistic, then the spirit of the majority of works of art is to be also optimistic, etc.

And, being in accordance with the dominating social ideas, art can influence its perceivers by *strengthening* their *social attitudes*, by adding a «portion of harmony» to their psychic life. Such is the *character of the contribution of art* to the main features of social life.

2. As to the *quantitative estimation* for this contribution, it may evidently change from one society to another, and even for the society studied this contribution may be different for different epochs. Nevertheless, it is desirable to know at least the *order of magnitude* of this contribution. For this purpose, quantitative measurements, concerning members of some concrete societies, are suitable. Such investigations belong to the field of *sociological measurements*.

To measure the impact of art, different sociological procedures may be used (see, e.g., Sokolov, 1990), but two main difficulties are inherent to all of them:

- it is almost impossible to separate the influences of art (or of different kinds of art) from many other influences;
- it is almost impossible to determine the causal relationship between the variables studied.

That is why the results of numerous previous investigations devoted to the influence of art could not be considered as quantitative ones. Only in recent years it did become possible to derive a procedure of measurement that permitted us to *eliminate collateral links* between the phenomenon studied and many other obstacles influencing the data used (Petrov, 1988). Further, we shall list several examples of the measurements resulting from the application of this procedure.

During a sociological investigation (done in Russia in 1981–1982) devoted to the cultural needs of the population (a fragment of the results was used, namely the data on the population of the city of Kostroma), a set of 20 indicators was studied concerning different aspects of the «*socio-moral potential*» of personality: views on material and spiritual values, social activity, family relations, etc. These indicators were all aggregated into an index characterizing each person (respondent) studied. (A method of aggregation was used based on a structural model describing how a person is accustomed to a dominating social value — see Petrov and Jablonsky, 1980; Suna and Petrov, 1985. This method gave results measured in a ratio scale, i.e., the indices obtained could be treated as usual numbers.) Then the *links between the index and the different kinds of contacts with art* were measured; the influences (on these links) of 22 variables were eliminated; these eliminated parameters included sex, age, educational level of the person, educational level of the person's parents, and intensity of contacts with different kinds of art (besides the kind of art studied). The procedure used permitted us to determine (though not in all the cases) the *direction of causal relationship* between the index and the intensity of contacts with the kind of art studied (e.g., to find out whether changes of the index cause changes in intensity of contacts with art or, vice versa, contacts with art influence upon the index).

The results obtained are indicative of a rather *strong influence of art* on the «socio-moral potential» of personality. Thus, each regular — additional per year — contact with theatre results in a 13% growth of the above index (here we mean, for instance, the difference between those who attend theatre twice a year and those who attend three times a year, between 5-time visitors and 6-time visitors, etc.). The causal direction of this link is from contacts with the theatre — to the index studied, i.e., it is theatre that causes such changes in the internal world of personality. Another example: each regular — additional per year — visit to a variety show is accompanied by 9% growth of the above index, but here we can observe no causality (probably, this case responds to more or less symmetrical interaction between contacts with art and changes of the index). Analogous results were obtained also for contacts with fine arts, art on TV and radio, reading books, etc. (In all these cases, we speak only about the average values statistically characterizing the overwhelming majority of the members of society.)

Now we can summarize all the results of the above type, taking into account the scales of different kinds of contacts with art, averaged over the population of Russia (e.g., intensity of theatre-going is near .6 per year, variety art concert-going — near .3 per year, etc.). Besides, one has to add so-called «indirect influences» of art (Suna and Petrov, 1985; pp. 235–258), i.e., the influences exerted by recipients of art via various channels of intrapersonal communications (in separate cases these «indirect influences» are 2–3 times higher than the direct ones). As a result, we can obtain an estimate of the average contribution of art (as a whole) to the «socio-moral potential» of personality; this estimate is close to 50%. In other words, approximately *half of the existing social (mental) stability* of society is a result of contacts with art in its different kinds. Such is the role of the «*emotional harmony*» carried by works of art.

And what might be our appraisal of this harmonizing of emotions to which art contributes? Naturally, this appraisal has to depend on our entire humanistic appraisal of the social stability to which art makes its contribution. But one has to remember that this contribution is made not directly to the social sphere; it first affects the sphere's individual elements, i.e., humans' souls. So we have to consider the *two faces* of this influence, the first of them connected with elements of the system each taken separately (*A*), the second one dealing with the system as a whole (*B*).

A. Undoubtedly, in all cases each *human internal world* is of unquestionable value, and the help in harmonizing this structure by means of emotional impulses (carried by works of art) is worthy of positive appraisal. From this point of view, every harmonizing impulse has to be evaluated as positive, even if it harmonizes the internal world both of a martyr (who is martyrized because of his noble deeds) and a murderer (who hates a martyr for these noble deeds). In the first case, such an appraisal contradicts neither common sense nor system regularities: to be good to a martyr, at least does not hurt any of the other members of society. But, in the second case, the situation is not so clear, and an evident contradiction takes place: harmony within the internal world of a murderer at the same time makes him feel confident, and that in turn probably permits him to commit new crimes, and we cannot evaluate such contribution as positive. However, here we actually deal with the second case of the influence studied, and so we turn to the following case.

B. From the point of view of the *social system* as a whole, its *harmonizing* means to facilitate the execution of the system's main social purposes. If these purposes are evaluated as positive, then the appropriate contribution of art may be considered positive. But, in other cases, the evaluation may be quite opposite. For instance, official art in Nazi Germany was in aid of inhuman activity. The situation in Russia during the last 70 years (and especially in the 1930s–1940s) was quite similar; the main aim of the ruling party was to stay in power, and so they needed to keep a stable social system, perpetuated by a low standard of living, mass repressions, etc. And all such actions could be realized *only through the instrumentality of the* above *socio-moral potential*, giving the people the possibility «to look ahead with confidence». So art played a rather important role in all these actions, strengthening the regime by means of specific emotional impulses.

There may be different ways to estimate the «price» of social progress or its absence (see, e.g., Petrov, 1989). However, one measure seems to be more or less adequate for the concrete problem studied: we have to calculate the number of *first-hand victims* of the regime, i.e., directly killed by it, and to take aside — of course, conditionally — the contributions of different social subsystems.

If we assume that the «weight» (role) of art in keeping social stability was approximately constant (at least in the order of magnitude), then about half this stability can be attributed to the impact of art. (And also about a half of this stability can be attributed to other forms of ideological influences: direct propaganda of appropriate views, their substantiation by dominating directions of official social sciences, etc.) Then the amount of direct victims attributed to this impact of art can be estimated as half the appropriate total amount during all the time of actions of the regime.

There are different estimates of the number of victims of the regime, some of them counting up to 111 million («Tri tsifry»... 1990), but for our rather approximate calculations, a number of at least 40 million seems to be about right for a time interval of about 20 years (1930s–1940s), which corresponds to 2 million per year. (This value is close to the estimate of the delay of social progress in one year, arrived at the basis of calculations of indirect consequences of the regime's functioning; the appropriate estimated number is near 1.5 million lives — see Petrov, 1989.) Half of these deaths — i.e., 20 million — can be attributed to the impact of art on social stability. Further, rather approximate calculations can be based on dividing this contribution among different kinds of art. Taking as a base, current intensities (scales) of contacts with different kinds of art and the appropriate «strengths» of their influences (Petrov, 1988), one can attribute, e.g., to literature at least 30% of the total contribution of art, i.e., about $20 * .3 = 6$ million. How can we use this value to derive an estimate for separate authors (writers, composers, painters, etc.)?

There exist certain grounds to suppose that each of the branches studied (literature and contacts with it, music and contacts with it, etc.) forms an integrity which possesses quite definite self-organizing regularities of its functioning. One of the most important of such regularities is the so-called Zipf's law (see also Section 1.4) that was named «the natural law for social sciences» (Petrov and Jablonsky, 1980). This law describes a quite inhomogeneous distribution of elements within any integrity of the self-organizing type: several of the most widespread names (elements) correspond to the overwhelming majority of the elements of the system. (For instance, in every language a rather small amount of the most widespread words cover the majority of all the words used; for example, in Russian the most frequently used words «в (во)» and «и» cover, respectively, about 4.3% and 3.6% of all the words of the current speech [Chastotny slovar russkogo jazyka, 1977; p. 807].) So, when speaking of literature and its functioning, one may take as elements the impulses resulting from contacts with different authors' works. And, if we assume a typical Zipf's distribution (and there are some statistical data proving this assumption), the most prominent authors will each cover about 3–4% of the total amount of impulses, which correspond to 180–240 thousand victims. Such is the estimate of the creativity of authors of the first ranks of popularity (like A.M. Gorky, A.N. Tolstoy, I.A. Dunajewsky, and others), who helped the regime by means of *emotional-harmonizing influences*.

Of course, the estimates obtained give only an order of magnitude of the influences studied. However, these estimates fit rather well in the entire system of knowledge concerning such matters (see, e.g., Petrov, 1989). Besides, such estimates may serve as a means of orientation or a warning for contemporary writers, composers, artists, etc.

4.6 The «Hamburg score» of art

So, works of art have rather different impacts on the spiritual life both of the individual and of society as a whole. That is why there exists a set of various *estimations* of works of art, and the criteria of the estimations are based on these different kinds of impacts. Which of these estimations are «true» — and which are «false», from the scientific point of view? We consider this problem from the point of view of our approach, which is based on the information theory. The results obtained appear to be important for many purposes, including the scientific foundations of art criticism, cultural politics, prognosticating, etc.

Starting from the point of view of our approach, the importance of any object for any system (that includes this object) depends mainly on the *effectiveness* of realizing the *functions* that are *inherent* to the sphere to which the given object belongs. This means that the impact of *works of art* has to be estimated in accordance with the effectiveness of their specific *«perfecting» influences*, common for all the objects that are part of the aesthetic sphere, i.e., in accordance with their ability to make the working of the human information system more perfect. (In this relation, the objects of the aesthetic sphere cannot be replaced by any objects belonging to other spheres.) Any other estimations seem to be incompatible with the contemporary system views, because such estimations can describe only the degree of closeness of the objects considered to some spheres for which the impact of art is not specific, «typical».

Taking this into account, let us look at the criteria that are usually applied both by scientists and art critics when estimating works of art.

The most widespread criteria are based on the *first-hand individual effects* of art, i.e., the effects directly accompanying perception and directly observed (or self-observed), being located «on the surface» of the processes of perception. These first-hand effects belong to everyday common sense, but sometimes they dominate even in the judgements of art critics. What phenomena are seen «on the surface» and therefore can be used as the foundations for the criteria considered? Two sides of the human spiritual life are evident to each observer:

— The *cognitive* aspect; the appropriate criterion consists in how much *information about the surroundings* is presented by a given work of art. And according to such criterion, the estimations of some people (sometimes even art critics among them) are higher if the given work of art provides a recipient with a rather rich information about reality (such situation responds to the concept of art as a «school of life»).

— The *emotional* aspect; the appropriate criterion consists in how much *pleasure*, or *positive emotions*, are connected with the perception of a given work of art. According to such criterion, the estimations of some people are higher if the given work of art provides a recipient with rather strong positive emotions (this concept responds to the hedonic views).

Both of the above variants of first-hand individual effects are so evident that it seems natural to take them into account. But, from the point of view of our approach, the above variants of estimations are, of course, insufficient. To *provide information about the outer world*, is a function inherent to objects of such «cognitive» spheres as science, education, interpersonal communications, etc. So the direct cognitive criterion (concerned with information about the outer world) can give only an estimation about how close this work of art is to objects of science, education, communicative means, etc. And to *carry pleasure* is a function inherent to many «material» objects (e.g., sweets). So the direct emotional criterion (concerned with the amount of positive emotions) also gives an estimation of some non-specific properties of a given work of art. As a result, both types of criteria, based on first-hand effects, do not concern the specific «perfecting» properties of art objects and would be rejected from the scientific point of view.

Quite opposite is the next type of criteria which use as their base the *resulting social effects* of art, i.e., the «remote» sequences of the processes of perception concerned with the «genuine sphere of social relations», including ideology, politics, etc. (One such estimation was described above; see also Petrov, 1991/Ref. 8.) However, again such a kind of estimation concerns properties that are non-specific for the aesthetic sphere, because the «purely social» effect is usually obtained by means of the appropriate objects being used for a particular social activity — such as, for instance, political propaganda, official laws, etc. And if the given work of art is valued from such a «sociological» point of view — this is nothing else but a measure of the closeness of this work of art to the objects belonging to the above spheres of social activity.

All the above confirms that it is unreasonable to search for our scientific criteria through the application of the usual motivations of art criticism, common sense, or sociological considerations: in artillery terms, we had either shot around or shot over the target. Evidently, the truth is somewhere in the middle, and possibly on a completely *different level*, i.e., on the level, or in the framework of, our *approach*, which is *based on the information theory*. Namely, the criteria relating to the specific impact of art (i.e., the impact that is specific for the aesthetic sphere in general) satisfy the scientific requirements. That is why we have to assume the *«perfecting» influence* of art — i.e., a special «training» of the human information system, assisting it to function more perfectly — as the *basis for* our *criteria*. (This concept is in agreement with the ideas of a famous Russian psychologist, Lew Vygotsky, 1971, who considered art a special sort of «emotional training».)

Works of different kinds of art, genres, and directions realize *different kinds of* such a «training» — like different kinds of sports train different functions of the human organism (e.g., rowing develops mainly the function of the lungs, skiing trains primarily the leg muscles, and some kinds of sports develop a set of functions). In the case of the human information system, perception of works of art can be used for rather specific kinds of training. As was mentioned earlier, these specific kinds are:

- training in the *processes* that are localized *within a given level* of the human structure of information processing and operate with *concrete object images*; this kind of training is «purely» (more or less) realized by art objects belonging to figurative painting and sculpture, dance, prose and verse;
- training in the *processes* that are also localized *within a given level*, but operate with *abstract features* of objects; this kind of training is «purely» realized by art objects belonging to geometrical abstract painting, ornament, colour music, «light» and «classical» music;
- training in the *processes* of *interlevel communications*, operating with concrete images or/and with abstract features, i.e., making a «translation» from one set of signs to another; this kind of training is «purely» realized by the *synthesis of different perfecting structures*, existing in the same art object (in other words, by the interaction between these perfecting structures), e.g., in song, the theatre, cinema, etc.

Naturally, in reality most art objects carry not only one of the above kinds of training, but maybe two or three; so the examples above concern only the main «special perfecting destination» of the appropriate objects. (As was shown earlier, in principle it is possible to make a more detailed division of the «perfecting training» and the responding objects, but now it seems unreasonable to take such a detailed nomenclature into account, because all the varieties obtained are nothing else but different realizations of the above three principal components of the «perfecting influence» of art.)

These three kinds of training do influence our psychic activity not only quantitatively (i.e., increasing definite abilities that already exist), but mainly *qualitatively* — as rather specific «tools» permitting us to create new views on the world, new attitudes, new relations, etc. Due to the above training, the functioning of the human information system is changed in such a way that it seems as if *new functional organs* are created. This reminds us of the changes in a human being interacting with the material world, which took place after material tools were introduced; quite similarly, art may be considered a tool for working within the spiritual world (with images, features, ideas, etc.). Such a training permits *culture* to rule the spiritual life of a person, and if genetic information forms a human body (and some initial principles of psychic life), culture forms the human soul. And due to art, our spiritual life is a certain more or less *self-consistent* whole, with different elements that can be decoded (while experiencing the appropriate situations) in a «*culture-dictated*» manner, including sorrow when autumn leaves are dying, joy at looking at a beautiful woman, heart palpitations because of the morning mist, etc. These sensations are all dictated by culture through the means of art, which makes the spiritual life of each of us not only self-consistent, but also *compatible* with the spiritual worlds of other members of society. Such are the final results of the training realized by works of art.

It seems reasonable to base our scientific estimations of works of art on their *effectiveness* in the above *three kinds of training*. So, at least *two problems* arise immediately dealing, first, with the concept of «effectiveness» and, second, with the «three kinds».

The problem of «*effectiveness*» is rather complicated because it concerns the links between a work of art, and a perceiving subject whose information system is trained. As to the work of art, it is a given work, but as to the subject — he (she) may be, in principle, very different, and therefore the effectiveness of the training may differ within a rather wide range. For instance, a qualified reader can get many experiences while reading a poem, which results in the appropriate «perfecting» influences. But another reader who can hardly understand the language of the same poem gets almost nothing from it. That is why the effectiveness of each perfecting influence *depends on the* perceiving *subject*. So what kind of subject, or what *perceptual process*, do we have to consider in the framework of our model? This is the key point of the problem of effectiveness.

Here we have to take into account that the resulting, «remote» purpose of our consideration is to *compare* different works of art on their perfecting influences. But such a comparison means nothing else but to have some «*standards*» concerning the *conditions* of the processes considered. And the only possible (and quite natural) *standardized conditions* (of the perceptive process) respond to the «*ceiling*» (the summit) of the perceptual opportunities, i.e., the upper *maximal «revealing» of the structures* contained objectively in a given work of art. What does it mean? It means that, e.g., if in a given work of art three structures (or devices used to exert a perfecting influence) can be *found* by an investigator, all these structures (or devices) have to be taken into account when estimating this work of art, though some recipients can identify only one or two structures. (As for the above-mentioned identification made by an investigator, different ways are possible here to make the appropriate procedures more objective, including, e.g., purely statistical and psychological methods.) So, practically, the *investigator's perception* (rather than sociological data) has to be used as a basis for further estimations. (It is clear that such a basis seems to be rather subjective, but it is a task of an investigator to check the supposed structures by purely scientific means; the pathos of this passage concerns only the lack of meaning of the usual sociological studies of preferences or opinions in this field.) Thus it is now all about the features of the perceptual process, and therefore — about the problem of the «effectiveness».

As to the problem of the «*three kinds*», — it is rooted in the very fact that there *are* three kinds of training. That is why an estimation based on one of them may contradict the other one or two estimations. Then which estimation should be considered the main one for a given work of art?

The solution to this problem is rather simple. Quite analogous to the case of the *specific functions* of objects belonging to different spheres, so similar is the situation when considering the *specific kinds of perfecting*, or training, of the human information system. Here we also have to base our scientific estimations on that kind of *perfecting training* which is *specific for the class of objects* to which the given work of art belongs. So if, for instance, the main perfecting «destination» of the given object (work of art) is to train the processes of a given level and deal with concrete object images, so the *estimation* of this object has to be based on this «own» function of such objects, but not on some other properties (based on the capability of this object to train other types of the information processes). Exactly such must be, e.g., the estimation of a painting belonging to figurative fine art, and here we must neglect other possible criteria based, for instance, on combinations of abstract features that are present in the given painting. Such *specific criteria* are valid, because none of the above kinds of training can be replaced by another one (like a deficit of Vitamin A cannot be compensated by food containing Vitamin B, so each need of an organism has to be satisfied).

Now, when we are solving problems concerning both the subject of the perceptual process and the «main kind» of training to be considered, the situation looks rather simple: when evaluating the given concrete object (work of art), an investigator (or an art critic) has to take into account primarily the *capability* of this object in the field *of* its «*main perfecting training*». Of course, other «perfecting capabilities» (as well as other, «by-product» functions of this object, dealing with its non-specific, non-aesthetic influences) can also be taken into account, but only as «additions» to the main perfecting capability of this object.

(Such an evaluation seems to be profound, especially in light of the contemporary tendency towards «divergent» development of different spheres of human activity, and this tendency also concerns different kinds of art — as well as different genres and directions within each kind of art. So, different kinds of art, genres, and directions are separated from each other, specializing in their «own» kinds of training; therefore, they realize more and more «purely» their specific perfecting influences.)

As for the further *quantitative evaluation* of different parameters of perception — meaning concrete recipients (including possibly also concrete investigators) and the significance of the above kinds of perfecting properties for them — such an evaluation is possible in principle. It can be done with the help of some psychological experiments, the results of which can be processed by means of the so-called non-metric multidimensional scaling (for a review of the applications of this method in the studies of aesthetic perception see, e.g., Kamensky and Petrov, 1982).

* * *

Many different kinds of estimations of works of art exist and sometimes compete with each other. Some aestheticians or art critics evaluate works of art according to their capability to provoke definite emotions (pleasure, sexual excitement, etc.), some people take into account primarily definite cognitive or ideological impacts of art, and so on. Looking at the giant crowds of rock- or pop-music admirers, some people of commerce evaluate the appropriate musical phenomena according to «commercial» criteria; knowing about the high prices of all the paintings by the great masters, some critics (belonging to the conceptual wing of art) declare that «only a touch of a master — but not a structure of a painting — contributes to the price of an object of art», etc. A panoply of various criteria is around us.

But we dare to consider «true» criteria only those that have a definite scientific basis, and the above model is amongst such approaches. Therefore, the criteria concerning specific perfecting impacts of works of art should be a scientific basis for comparing different works of art and estimating them. And, according to this approach, the work of art cannot be «about something» in the outer world (e.g., «about love», «about war», «about crime», etc.), but only *«about the internal structures»* of information processing which need a support of art. So, what is the intensity of this support — such must be the estimation of the given work of art.

Once upon a time there existed a legend about the so-called *«Hamburg score»*. Usually sportsmen — especially boxers — while their competitions do not reveal their real possibilities, and behave according to definite rules, the nature of which is rather far from the specificity of the appropriate kind of sport (e.g., a winner is defined depending on which victory could give the maximal benefit to the manager). Only once a year the criteria, which are specific for this kind of sport, are the decisive ones: only once a year in Hamburg a secret (for the public especially) competition of the strongest sportsmen in the world is held, and this competition provides them with the only opportunity to *determine honestly* which of the sportsmen are the *strongest* ones — according to the *«intrinsic»* sport *criteria*. Such estimations are called «Hamburg score». Exactly such a «Hamburg score» for works of art was proposed above in the framework of our approach.

Chapter 5
Devices of art

Ainsi le choeur des romances
A la lèvre vole-t-il
Exclus-en si tu commences
Le réel parce-que vil

Le sens trop précis rature
Ta vague littérature.

Stéphane Mallarmé, Toute l'âme resumé...

Art is a means of training and perfecting the human ability of informational interaction, a means of «education of the senses». On the other hand, art is the best laboratory where man perfects and finishes various devices and methods of informational interaction. As we know, this interaction passes through three stages (see Chapter 2). Correspondingly, these devices have three main purposes:

1. To *attract* the attention, to establish «informational contact», informational interaction between the work of art and the spectator (reader, listener). That is, to pass the first stage, to overcome the threshold, to break the routine, habitual notions and let the stimulus be perceived and the emotion positive. It is obvious that before the subject can aesthetically estimate the stimulus, he must perceive it, make it an object of his attention and consciousness. But how often do we look, but not perceive; read, but not understand; hear, but not realize! The first problem of the artist is *to force* a spectator to perceive!

2. The second problem is *to keep* the attention of the subject. This means — *to prolong* the second stage of perception and the positive emotion for as long as possible.

3. The last problem is — *to prepare* some new, other, contrast stimulus at the beginning of the third stage. Informativeness of the old stimulus is inevitably exhausted, the stimulus becomes uninformative, habitual, «boring». And the artist must provide the conditions that permit the subject to abandon the old stimulus, to switch to the contrast stimulus and begin a new circle of perception.

Fig. 5.1 a)

5.1 How to attract attention?

We know the conditions (2.20) for the threshold overcoming:

$$p_{ij} - p_{i0}p_{0j}/[p_{00} - I - \beta(r_{ij} - R)] = 0. \tag{5.1}$$

All devices and methods that increase the left side of this expression are exactly those means of art that can attract the attention and establish an informational intraction between the artist and the spectator (reader, listener).

Let us remember what these conditions and corresponding devices of art are (see also Chapter 2).

a. *Repetition*: if any stimulus x_i is repeated, then the first exposition of it facilitates the perception of the second one. Thus, the simple repetition and amplification of a stimulus is a device to overcome the threshold. *Symmetry, meter, refrain, leit-motif, reminiscence, comparison, allusion, etc.* are examples of such devices. In imitative arts (painting, literature, etc.), the natural object can play the role of the first stimulus, the image of this object — the role of the second one. Hence, *realism*, the imitation of the real object by its image, facilitates the perception of the last one. The real image is perceived more easily than the abstract one — here the well-known effect of *recognition* is at work. Hence, *realism* in imitative arts can be considered to be one of the aesthetic devices.

b. *Concretization*: the more «rich», concrete, exact the stimulus, the more features are associated with it, the easier it overcomes the threshold. This device is closely connected with the previous one, because both these devices are intended to increase the growth of the probability p_{ij}.

Concretization is one of the most important aesthetic devices; and concreteness is one of the main properties of art that distinguishes it from other forms of cognition such as science, philosophy, etc. Some examples of consequent processes of concretization in graphic arts are presented in Fig. 5.1.

In literature, concretization is expressed through the addition of new details and features to the image described. Such devices as comparison, epithet, and metaphor are means of concretization. The sentence «A *red* cat has appeared on the stairs» will be more concrete, informative, and aesthetic than the simple «A cat has appeared on the stairs». If the probability of the red colour among cats is,

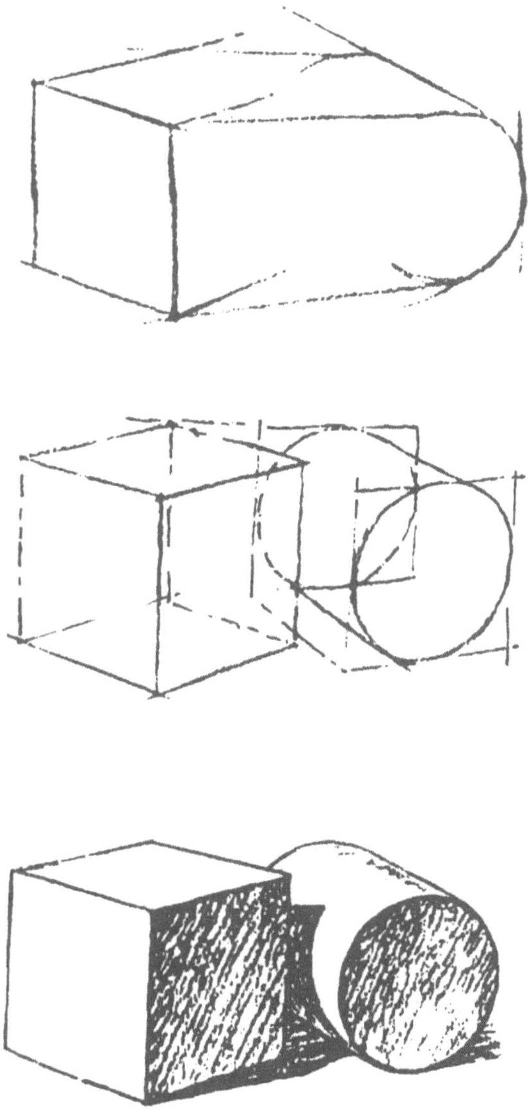

Fig. 5.1 b)

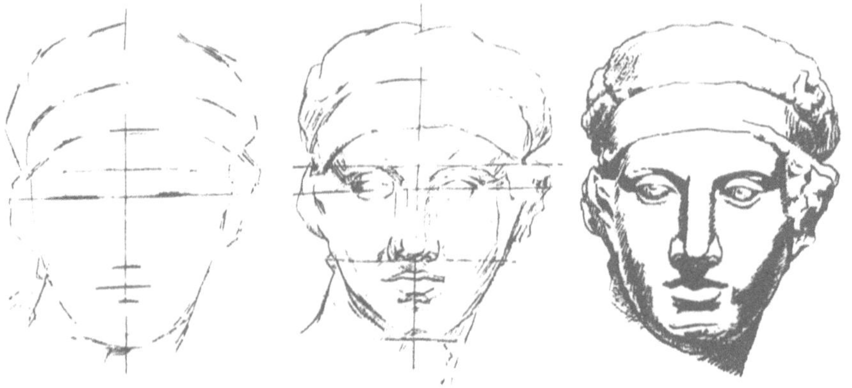

Fig. 5.1 c)

for instance, $p(red) = 1/32$, then the epithet *red* brings us $\log_2(32) = 5$ bit of information!

A Russian writer, Grigorovitch (1964), remembers with admiration a lesson in literature mastery he got from Dostoevsky: «No, no!» — Dostoevsky said irritatedly — «you make it too dry: a penny fell at his feet. You must write: a penny fell on the pavement, rang and jumped up.» Two concrete details transform an abstract notion into a visible and audible image.

c. *Freedom from prejudice*: one should be free of clichéd thinking and stereotypical perceptions when this stimulus x_i is rigidly connected with the other response y_0 and vice versa, so that «mixed» probabilities p_{i0}, p_{0j} are too large. In other words: the ability to perceive, unencumbered by perceptions, something new.

d. *Contrast*: the stimulus x_i must be accompanied — in space or time — by another, contrast stimulus x_0; this decreases the threshold and is a necessary condition of perception in general: the stimulus can be perceived only with the contrast stimulus in the background.

e. *Decrease the informativeness I of the nearest background*, do away with other interesting, bright, informative stimuli, that can compete with x_i. For instance: there must be no disturbing sounds and conversations during a concert or performance; a picture must be positioned on a neutral, grey wall at adequate distance from other pictures; bright images in a work of literature must be interrupted by relatively neutral material, etc.

f. *Removal of any physical obstacles* to facilitate perception of the stimulus and decrease the resource expenses r_{ij}, which are necessary to perceive the stimulus. Obvious devices of this kind are placing the work of art in a central, well-observed position, and providing adequate light, good audibility, etc.

g. *Economy of the resources* and decrease in the deficit of the resources β. As is obvious from Fig. 2.2b, a decrease of β can solve two problems: (1) to lower the threshold; (2) to raise the optimum; and, as a consequence, to enlarge the longitude of the second stage and the positive emotion. That is why economy of resources — time, space, energy, number of operations and elements (words, paints etc.) — is the foundation of many aesthetic devices. In literature, ellipsis is an example of these devices: words that do not bring new information can be omitted. In pictures and paintings silhouette or contour are often sufficient, and other details can be omitted and then restored by the spectator, etc.

The capacity of the spectator for «co-creation» is one of the ways the artist can omit the more obvious and, consequently, less informative details, and thus to economize on the resources. It also provides an increase of effectiveness E and an additional aesthetic pleasure. It's very important to create the most *effective* image, which can win the struggle with other images for the attention of the spectator (reader, listener). Let us suppose that the artist has found the most informative details and features of the image. Still, this image could not overcome the threshold of consciousness if these details would be dissipated, divided one from another in time and space.

This fact is especially obvious in literature. The reason is that one word usually transmits only one abstract feature of an object. Many words are needed to create a whole, concrete image of this object. Specific difficulties of forming the image by means of words were already noticed by Lessing (1955) in *Laocoön*: «Things which the eye envelops at once, a poet must show us slowly, in parts. And often we have forgotten the first part when perceiving the last one.»

Hence, the writer must not only increase the informativeness of features, but decrease the resource expense — time, space, number of words, etc., which are used to transmit these features. It would be ideal to present all the features of the object simultaneously. But literature is the art of the sequential exposition and simultaneity in the exact sense is unattainable here. It is still desirable to approximate this simultaneity as close as possible by the compression of information and the elimination of all superfluous details and features, which do not bring new information, but merely demand resource expense.

The capacity of the reader for co-creation permits the artist to omit these details. He must only have faith in the reader and, of course, be able to select the most informative details, which is one aspect of his talent. Let us remember the famous monologue of Treplev from Chekhov's «Sea-gull» (1957): «It is easy for Trigorin, he has devices: the neck of the broken bottle sparks on the weir and the shadow of the water-mill wheel darkens — and the moon night is finished. And me? There are the flickering lights, and the quiet star twinkling, and the distant sounds of piano dying away in the silent aromatic air... It is poignant!»

The importance of the economy of resources in art is affirmed by many facts. We know the watchful attitude of sensitive artists against any temptations that would allow an uncontrolled extension of expressive means of art. Such extension could only decrease the effectiveness of art and the aesthetic pleasure caused by it, if not accompanied by the corresponding increase in content (information).

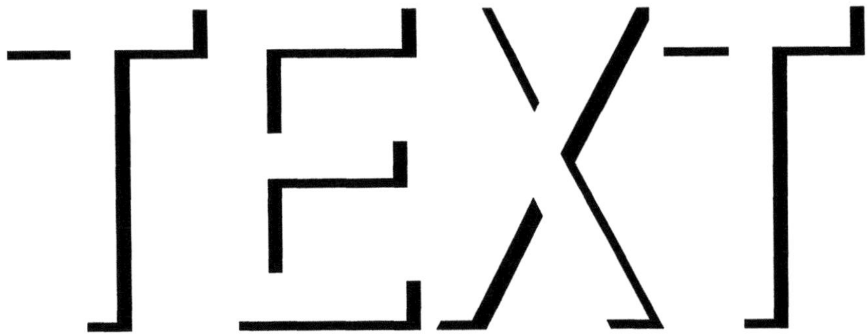

Fig. 5.2

The history of art gives us many examples of conscious self-restraint when using expressive means of art. Ancient Greek painters needed only four paints: black, white, red, and yellow. Any attempt to expand this spectrum was treated as «barbarism». Also, it is well known that some actors and producers were against the introduction of sound — as well as colour — into the art of cinema. «Brevity is the sister of talent» — this Chekhov formula is true for all the arts.

5.2 How to keep attention? Nuance

The subject overcomes the first compulsory stage under the influence of the artist and self-compulsion. But in the second stage we observe compulsory, as well as spontaneous (free) behaviour.

We know that if the combination (x_i, y_j) has overcome the threshold, then the probability $p(x_i, y_j)$ can increase spontaneously (see Chapter 2). This spontaneous behaviour forms the basis of the so-called *co-creation*.

The effect of co-creation in art is well known. This effect plays a significant role in the perception of a work of art and gives additional aesthetic pleasure to the spectator (reader, listener). Co-creation permits the artist to omit the less informative details of the image and to economize resources of energy, time, number of operations, etc. It increases the effectiveness of the work of art and becomes the basis of many aesthetic devices.

Many such devices are based on such a feature of the image, as the ability of the whole to restore some absent details and features. The simplest example is demonstrated in Fig. 5.2. The word «TEXT» is written there. But is it really written? The left contours of the letters are absent, aren't they? Yes, reader, but your imagination reconstructs the details lacking and allows you to perceive the letter correctly as a whole.

This phenomenon is known in physiology, psychology, and philosophy under various names: assimilation, apperception, dominance, idealization, aspiration to the ideal form, etc. We have named it idealization. The essence of this phenomenon

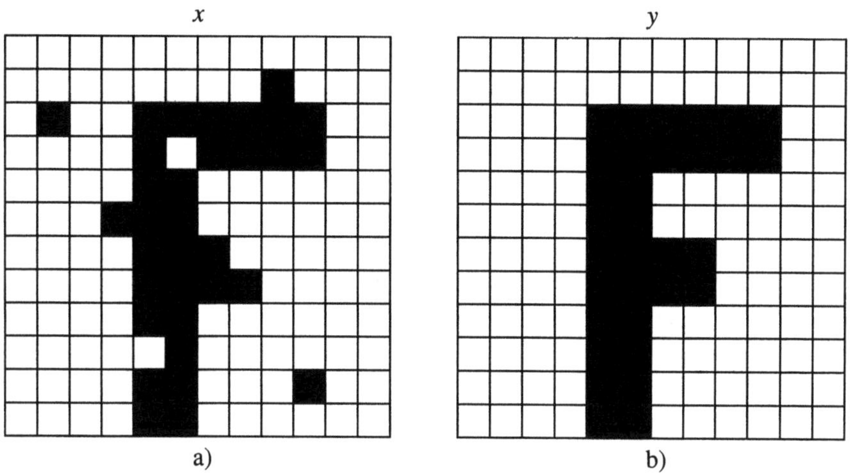

Fig. 5.3

is as follows: when any pattern (combination of features) overcomes the threshold, then the spontaneous movement to any ideal state begins. This state plays the role of an «attractor», which can attract and assimilate near patterns.

This attraction is a consequence of the principle of the information maximum; in this case it means the following: if the features of the stimulus $\{x^1, x^2, x^3, \ldots, x^n\}$ are fixed (given compulsory), then the subject chooses the response y that contains the maximum of mutual information about these features. When the response is chosen, it begins to exert a reverse influence on the features of the stimulus and the subject aspires to change these features into that direction, which provides a further increase in the mutual information.

The possibility for such an increase is present when the stimulus contains some errors, noise, lacks certain details, or has superfluous ones, etc., i.e., if some relation derivates from the «ideal» pattern (see Fig. 5.3). Any error diminishes the mutual information between x and y. And correction of the error leads to an increase in this information. This correction becomes possible if the compulsion from the stimulus is weakened (weak light, quiet sound, high level of noise, distracted or scattered attention, etc). The picture in the raster X (Fig. 5.3a) will be perceived as the letter «F» (Fig. 5.3b), though some black cells are absent and some others are superfluous.

It was suggested above that positive emotions arise only in free, spontaneous behaviour, i.e., in the process of co-creation. Hence, co-creation is the main source of aesthetic pleasure and a necessary condition for any aesthetic perception.

This is the reason why a too «perfect», finished work of art looks so «dull»: it forces the spectator (reader, listener) into the ideal state and deprives him of the opportunity of free behaviour and the pleasure that accompanies this behaviour.

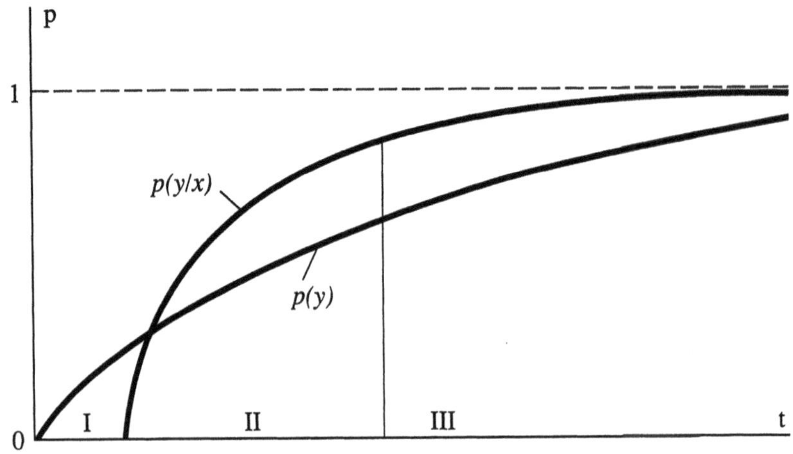

Fig. 5.4

The purpose of the artist is to help the subject overcome the threshold. The subject can execute further moves spontaneously, under the influence of the attractive forces, which lead him to the ideal state. Emotions are exactly such forces.

But this does not mean that from this moment on the artist can leave the spectator for a long time. No, the informativeness of any stimulus is sooner or later exhausted. If the artist occasionally does not give something new to the spectator, he will lose the informational contact with him.

The two main groups of aesthetical devices that permit the artist to keep the attention of the spectator, to prolong perception, are *nuance* and *contrast*. Let us look at the first one in detail.

The simplest and most primitive means to attract and keep the attention fixed on some stimulus is repetition of this stimulus. Let x be the first stimulus, y — the second one. In the case of simple (full) repetition, all the features of these stimuli (except, of course, their positions in space or time) coincide: $x = y$. If these stimuli are repeated, then both probabilities $p(y/x)$, and $p(y)$ increase, but $p(x/y)$ increases more rapidly than $p(y)$. After overcoming the threshold, the mutual information between x and y

$$I(x,y) = \log \frac{p(y/x)}{p(y)} \tag{5.2}$$

at first grows (the second stage). But later, probability $p(x/y)$ approaches 1 and stops growing. The information begins to decrease with the growth of $p(y)$ (the third stage, see Fig. 5.4). It means that a simple repetition becomes «boring», non-informative.

It is possible to prolong the second stage, to increase the information and the positive emotion. If one begins to *vary* variables x and y, but *conserves the*

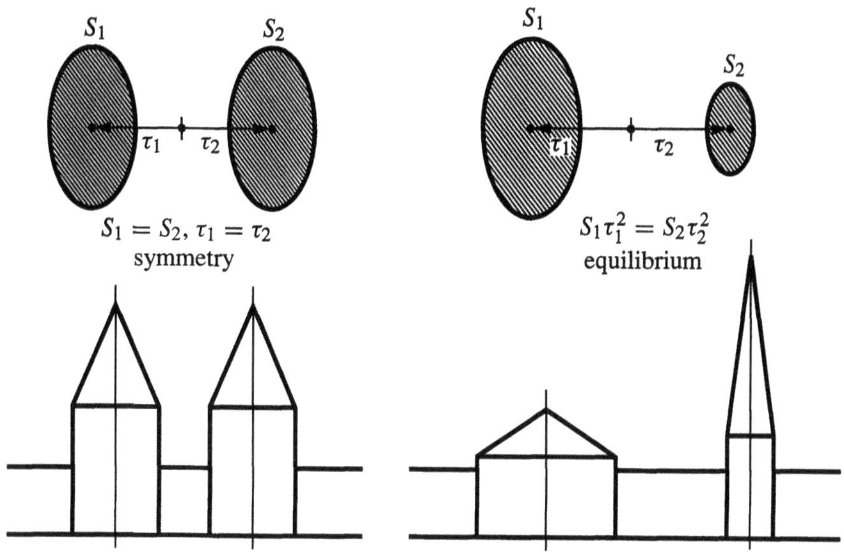

Fig. 5.5

connection between them, then probability $p(y)$ decreases, $p(y/x)$ remains the same and information $I(x,y)$ continues to grow. The numerous devices that provide such a development are known under the names *nuance, variation*, or *partial repetition.*

Partial repetition (when only part of the features or their combinations are repeated) can be considered as a further development of the full (simple) one. So *equilibrium* in painting and architecture is a further development of simple *symmetry.* In this case, areas or volumes must not be equal, but *combinations* of them (as the product of the area to its distance from the centre, Fig. 5.5) can be. Other examples of partial repetition are *similarity* and *proportion* in architecture (Fig. 5.6), and *transposition* in music. In this case, the combination repeated is a *relation* of features (lengths, frequences), but not features themselves. These devices can be considered as a development of *identity* when the features themselves are repeated.

Many geometrical and topological transformations change one part of the figure's features and preserve other features or their combinations («invariants»). For example, a curve remains a straight line, circle, parabola, etc., and only its parameters (inclination, radius, crookedness, etc.) vary. Folds in clothes, lines of waves, rays of light, and other examples show us such transformations and their invariants in ornament, design, graphic arts, etc. (Figs. 5.7a, b, c, d).

In poetry we find such examples of partial repetition as rhythm (coincidence of a part of the accents) and rhyme (coincidence of a part of the phonemes). It is

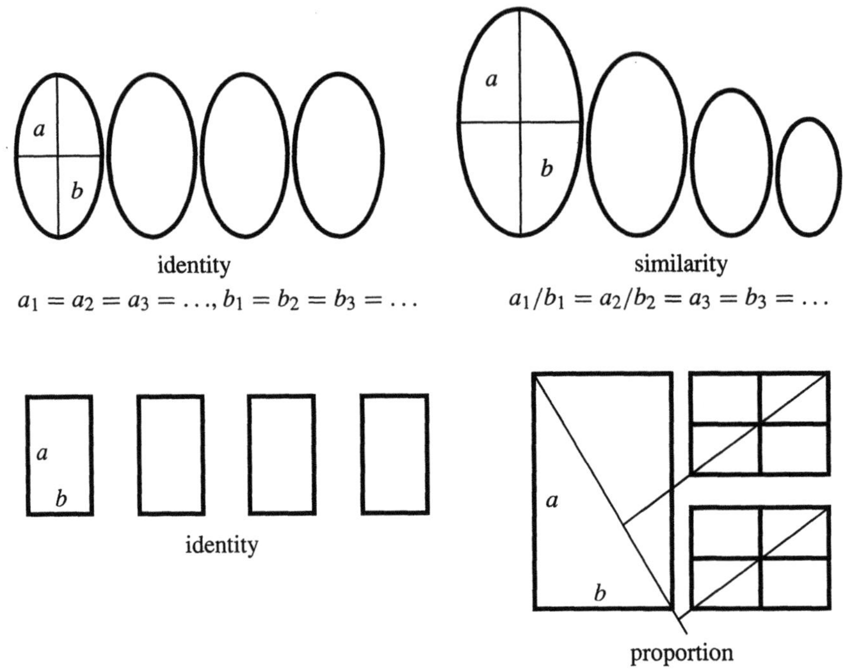

Fig. 5.6

important to emphasize the word «part»: lack of coincidence destroys rhythm and rhyme, full coincidence impoverishes them.

In music we find such examples of partial repetition as *consonance* — coincidence of a part of overtones; *transposition* — when absolute altitudes of sounds change, but intervals between them are preserved; *variation* and *elaboration* of musical theme, etc.

In architecture and music we find such a device as a *module*, based on partial repetition. A module is a common unit of length or longitude for spatial or temporal segments, used in a work of art. Lengths or longitudes of these segments must be divisible — without remainder — by this unit. It allows variety and, at the same time, provides a unity of spatial or temporal relations in the work of art. The *time* (or part of it) is the example of a module in music. It provides the unity of metrical and rhythmical correlations in music. The *step* of division is an example of a module in architecture.

For many works of art, the title or name plays the role of the «first stimulus» x. The title provides us with a general notion about the content, and we perceive all other information that the work of art gives us («second stimulus» y) with this notion in the background. Hence, the choice of an exact title plays a very important role in the perception of a work of art. A good title must not repeat the content, but has to have a relation of «partial coincidence» with it.

Fig. 5.7 a)

It is necessary to say that it is a common mistake of many artists to pay little attention to the titles of their works. Usually a title only repeats what a spectator can see, without contributing anything new. Of course, we cannot demand that each artist be a poet and can control the word. But it would be well to aspire to this. Undoubtedly, the title adds a lot to the charm of the «Moonlight Sonata». It gives a certain direction to the senses of the listener, a certain image so he/she can perceive the music with this image in the background. Alas, we know this title proposal did not come from Beethoven.

Another example: the famous producer N. Akimov, in his book, «About Theatre» (1962) expresses interesting thoughts about partial repetition (in quotes — our commentary):

«A character of a comedy («second stimulus») must reflect his/her prototype in life («first stimulus»), but not be an exact copy of it... In this question... one must stay within reasonable limits. If we simply transfer a living prototype to a comedy, the image will be insipid; if the exaggeration is too large — the connec-

Fig. 5.7 b)

tion between the personage and the prototype will be lost — a very dangerous phenomenon of inexpressive grotesque».

«It seems to me that the attention the spectator directs towards the scene is conditioned by two contrary things: the things every spectator has in common with the stage action, and the distinction between his experience and what is happening on the scene... The equilibrium of these two factors is the secret of the attractiveness of a work. When the first decreases and the spectator does not find anything in common between his experience («first stimulus») and the stage action («second stimulus») he remains untouched. If there is a form of life on Mars that sharply differs from ours, it would be impossible for us to bear a Martian performance... On the other hand, if a spectator finds nothing new, nothing that enriches his experience — he vainly wastes time in the theatre».

Now we want to compare full repetition with partial repetition. Both these devices can be used for overcoming the threshold, as well as for prolongation of the second stage of perception. But it is profitable to use the full repetition

Fig. 5.7 c)

Fig. 5.7 d)

during the first stage and the partial repetition during the second one. There is the well-known psychological fact that there are features that are perceived and recognized at first sight, more easily than others; and features that can attract and keep attention for a longer time than others. We have grouped some of these features into two categories:

– contrast	– nuance
– identity	– similatity, proportions
– generality	– detail
– continuity	– discretion
– straight line	– curved line
– smooth contour	– broken contour
– metre	– rhythm
– symmetry	– equilibrium

We see that the left row is the manifestation of simple repetition and the right one — of partial repetition. On the other hand, it is a psychological fact that the «left» features are perceived and recognized earlier than the «right» ones. We perceive contrast before nuance, identity before proportions, generalized forms before details, and so forth. One may say that the features of the left column are more adequate to overcome the first stage, and the features of the right one — to prolong the second stage.

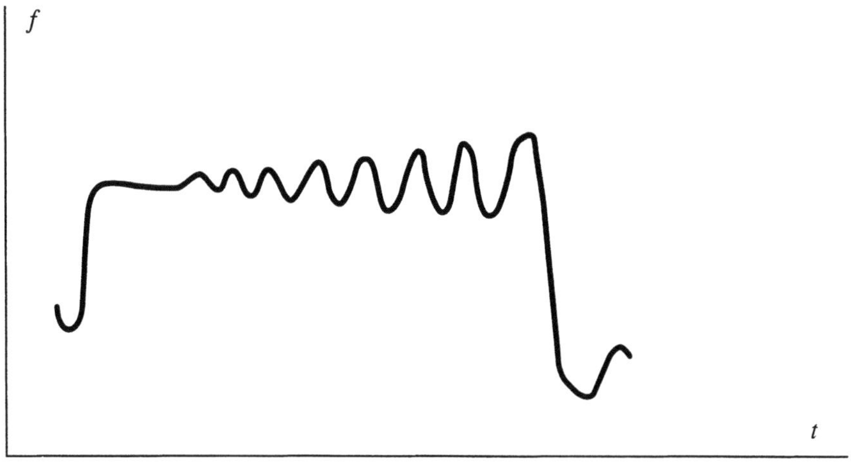

Fig. 5.8

A good illustration of these regularities is a musical device like the «vibrato» — slight oscillations of force, frequency, and other parameters of sound. The typical picture of the development of sound is shown in Fig. 5.8. The sound begins as «straight» and even (simple repetition), because it takes some time for a new sound to overcome the threshold, to be perceived and identified by the listener. But soon such simple continuation of sound becomes «boring», uninformative. Slight deviations of parameters of sound arise, the amplitude of these deviations increases, and at last the sound is replaced by another sound. Then the picture is repeated.

5.3 How to switch attention? Contrast

But sooner or later partial repetition and nuance variations of the stimulus exhaust their possibilities for maintaining informativeness of the stimulus and keeping the attention of the subject. Then a more radical change becomes inevitable: the subject abandons the old stimulus and begins to search for any essentially new, «other» stimulus. And he/she must find this stimulus in the framework of the work of art — otherwise he/she will find it somewhere else! In other words, he will turn away from the work of art.

The artist (writer, painter, etc.) must provide the subject with an opportunity for such searching by a special organization of the work of art. The means for such organization is *contrast*: each element of the work of art («stimulus») must be accompanied (in time or space) by another, contrast stimulus. In temporal arts (theatre, literature, music) the artist himself must alternate one stimulus with another one: tragic with comic, loud with silent, slow with quick, etc. In spatial arts (painting, architecture), he must give the spectator the opportunity to alternate

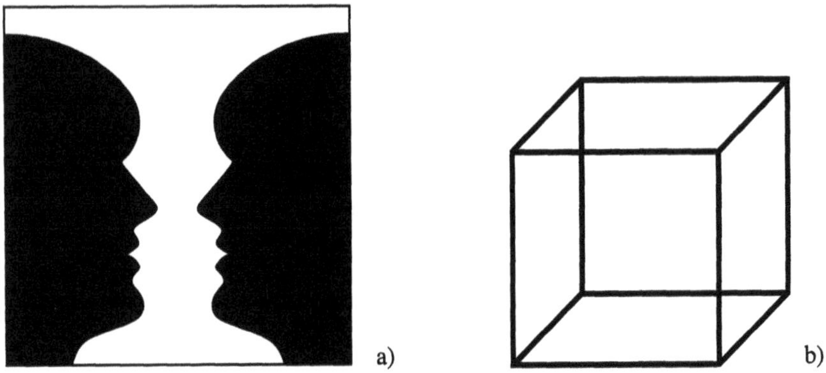

Fig. 5.9

this stimulus with the other contrasting one: dark with light, great with small, high with low, and so forth.

The phenomenon of contrast is a basis of many devices of art and regularities of the aesthetic perception that often far exceed the frames of art. (Some of these regularities are inherent even for animals.) Let us look at some of them.

a. *Oscillations of the behaviour.* If the stimulus x_i acts and the subject answers with response y_j, then probability p_{ij} increases. When this probability passes the point of the optimum, the emotion becomes negative and sooner or later the subject switches to another, contrast stimulus x_0 and its corresponding response y_0. The higher the probability p_{ij}, the lower the threshold for the combination (x_0, y_0). Then the switch happens, the subject passes into the state (x_0, y_0) and probability p_{00} begins to grow and p_{ij} — to diminish until a new switch occurs. And so forth.

Such oscillations are the ordinary state of all living beings and are observed on various levels of biological organization — from intracellular fluctuations of concentrations of chemical substances to the waves in quantity of the biological species. Human activity is not exception.

For example, if a rat can get food at both ends of a labyrinth, it will periodically switch from one end to another one without any apparent utilitarian cause. This fact was often observed in experiments. We have to assume that these oscillations are caused by aesthetic factors: the old becomes «boring» and the new — more attractive. (In these oscillations the new is often «the well-forgotten old»).

For our purposes, the most significant facts are the oscillations of the subject's attention and the subject's notions. In Fig. 5.9a one can see an ambiguous picture that can be perceived either as a white vase or as two black profiles. If you look at this picture long enough both images (responses) will alternate. In Fig. 5.9b you see the famous Necker's cube which you can look at either in the normal or in the inverse position.

These oscillations play an important role in aesthetic perception because they maintain the maximal level of aesthetic pleasure.

b. *Fixation of a point.* The greater the longitude or the area of the visual stimulus, the higher the probability (relative weight) of it. Both these factors can partially replace each other. A small area can be compensated by a long time of observation and vice versa to get the optimal value of probability.

Let us take a pair of contrast stimuli — a white point against a black background. We know (see above) that in this case, oscillations arise. The attention of the subject will switch from the point to the background and back. But the area of the point is much smaller than the area of the background. So the point perception time must be much longer than the background one. Theoretically, there are oscillations in the attention, but in practice the subject fixes his/her attention almost the whole time on the point and only sometimes switches to the background; i.e. main fixation is on the point.

This feature of the point is often used in psychology and in art when one wants to fix the attention of the subject on some area of the visual space. The point (or, generally speaking, the small contrast object) has the important feature «to keep attention» and, because of this, unites the complicated composition; «to be the center of relations» which can simplify the system of relations between elements of this composition.

There is a temporal analogue of this effect: the short temporal intervals, contrasting to the background (knocks, ticks, clicks, etc.), become the moments of concentration of attention.

c. *The Broca-Sultzer effect* (1902). There are optimal sizes (temporal or spatial) for the two contrast stimuli that provide a maximal aesthetic effect. The temporal effect consists in the following: there is an optimal longitude for each stimulus which provides the maximal subjective «force» (brightness, loudness, power) of the sensation and the maximal aesthetic pleasure. If one shortens or lenghtens the stimulus, this causes a decrease in the force of the sensation and aesthetic pleasure (Fig. 5.10). Thus, subjective brightness of a flashing light will be maximal when there is a certain relation of longitudes of light and dark periods.

This effect can be explained as a consequence of the three-stage law. The force of the subjective sensation is determined by the relative part the second stage (stage of amplification, see Fig. 2.4) plays in the total longitude of the stimulus. As one can see (Fig. 5.10), the relation is maximal if the end of the stimulus coincides with the end of the second stage. Both in shortness and in length of stimuli, the second stage occupies the lesser part of the total longitude. Hence the average amplification, which is perceived as the subjective «force» of the stimuli, will be less.

Besides a temporal effect there is a spatial Broca-Sultzer effect: the maximal subjective brightness of the stimulus is achieved when the optimal spatial size of the stimulus takes place. Both an increase and a decrease of this size cause a decrease in the subjective force of the sensation. For instance, a white spot against a black background will be the brightest if it is of average size. A spot which is too small will be assimilated by the background and will become invisible. On

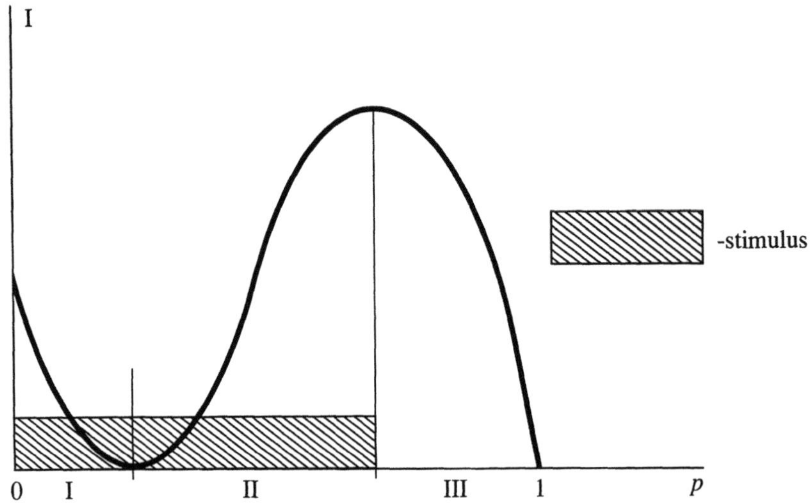

Fig. 5.10

the other hand, if the spot is of the same size as the background, the subjective brightness will be less because the emphasizing effect of the contrast disappears. The explanation of this type of effect is the same — the 3-stage law — because p_{ij} can be increased by a spatial extension of the stimulus.

For aesthetic perception, the Broca-Sultzer effect means that there are optimal sizes and durations of stimuli that provide a maximal aesthetic pleasure. And an artist must take into account this fact and use it to achieve the optimal organization of his work of art. There is an optimal distance between lines against a contrastive background, and there is an optimal time interval between knocks in a rhythm that make the rhythm the most effective and beautiful, etc.

d. *Boundary effect.* One of the most interesting consequences of the behaviour oscillations is the so-called «boundary effect». It is shown, in particular, in the movement of the eyes. If a picture consists of white and black spots, the eye moves mainly along the boundaries of these spots. The cause is obvious: near the boundary the switch from black to white and back again delivers a maximum of information (and positive emotions) with a minimal resource expense during the switch; i.e. effectiveness is maximal. This explains why the eye «loves» the contours of figures and avoids their internal areas.

In art, the boundary effect is used when a painter replaces a figure by its contour. This device can also be considered in the framework of a more general effect of «co-creation».

But the boundary effect is observed in many other forms of behaviour. For example, ethologists know well that the edge of the forest, the sea coast, the border of the ploughed field, and in general any peculiarity in the monotonous country attracts all living beings: birds, beasts, insects, etc. The population density near the boundary is always higher than in the inner areas. An important role here is played by a variety of conditions (informativeness!) promising a more interesting life.

People also prefer to settle near any «peculiarity»: the river, the sea coast, the boundary of a forest, etc. For proof, it is enough to look at a map depicting population density.

At least, if we look at how life in general is distributed on Earth, we will see that it «teems» with the boundaries of the three medias: the sea, the land, and the atmosphere. Moving away from the boundary deeper into each of these medias causes the pressure of life to weaken, and the concentration of biomass to decrease.

5.4 Laws of harmony: Golden section

A large amount of aesthetic devices must be united for the achievement of one goal: harmony of a whole. This harmony is the totality of some relations — spatial, temporal, logical, numerical, etc. — between the features of an aesthetic object. There exist some stable, constant values among these relations which provide a maximal aesthetic effect. Maybe the most famous one is the «golden section» which was already known in ancient times. The others are musical series constants.

The main principle of the harmonic organization of a work of art is the well-known principle, «unity of variety». Let us consider any two elements (spots, lines, sounds, etc.) of a work of art. Each of them can be described by means of some features (colour, length, frequency, etc.). Let us call these features x for the first element and y for the second one. Then the variety of the variable y is expressed by the entropy $H(Y)$, the unity between y and x — by the conditional entropy $H(Y/X)$ (with the sign minus), and the unity of variety can be described by means of mutual information between these variables:

$$I(X,Y) = H(Y) - H(Y/X). \tag{5.3}$$

According to the principle of the information maximum, the organization of a work of art is intended to achieve the maximum of this quantity. The more $H(Y)$, the more self-information (variety) variable y contains. The less $H(Y/X)$, the more the order and organization of the work of art, the easier to predict y if we know x.

But we know, it is also necessary to take into account the economy of resources $R(X,Y)$, so that a third item has to be supplied to the expression (5.3):

$$L(X,Y) = H(Y) - H(Y/X) - \beta R(X,Y) = \max. \tag{5.4}$$

And, the general principle of harmonic organization of the work of art, with accordance to this expression, is *«unity, variety, and economy»*.

One of the most interesting applications of the principle of the information maximum is the deduction of various psychological and aesthetic constants from the principle. An ability to deduce and, therefore, explain any fundamental constant is an important argument confirming the theory. So, one of the decisive arguments in favour of the electrodynamics by Maxwell was the possibility to theoretically deduce a well-known constant — the speed of light. Bohr's very strange and contadictory theory of the atom was accepted by the physicists, owing to the fact that the Ridberg constant could be obtained with a high exactness on the basis of this theory. Are there any such constants in psychology? Yes, we know, at least, one such constant in the psychology of perception: it is the *golden section.*

It is known that the golden section (golden proportion) is determined as such a relation between two quantities when the larger quantity x^0 relates to the lesser one x^1 as their sum relates to the large one:

$$\frac{x^0}{x^1} = \frac{x^0 + x^1}{x^0} \tag{5.5}$$

The numerical value of this relation is equal to approximately $1.618\ldots$.

From ancient times, the peculiar attractiveness of this proportion was known. Many painters, philosophers, psychologists, and mathematicians tried to explain this attractiveness, to ground this preference theoretically, to demonstrate that the golden proportion is — in any sense — the best, the optimal relation. But these attempts, in our opinion, had no adequate basis for this explanation. There was no criterion of optimality which could be related to the goals of human behaviour. If the principle of the information maximum is such a criterion, then the golden section (as well as the other psychological and aesthetical constants) has to be deduced from this principle.

Let us consider a situation where the resource R is proportional to the number m of operations, $R = am$, which is necessary to determine a relation of two quantities x^0/x^1. It can be two linear segments, two frequencies of musical sounds, two areas of spots, and so on. For clearness we shall look at a relation of two linear segments.

But how can we determine this relation x^0/x^1 if we cannot use any compasses, rulers, or other instruments and cannot transform these quantities into numbers? This situation is typical for the aesthetic perception!

Applying to the spatial segments, the procedure can look as follows: we compare segments, subtracting the lesser segment x^1 from the greater one x^0. If the number of subtractions is exactly m_1, the problem is solved and the relation $x^0/x^1 = m_1$. If there is some remainder x^2, such that $x^0 = m_1x^1 + x^2$, we must evaluate the size of the remainder. We can do this by repeating the same procedure, i.e., subtracting the remainder x^2 from the lesser segment x^1. The procedure must be repeated until the last remainder x^{n+1} becomes less than the setting error of measuring ϵ (Fig. 5.11). Then the last segment x^n can be treated as *module* μ — the common unity of measurement of two segments.

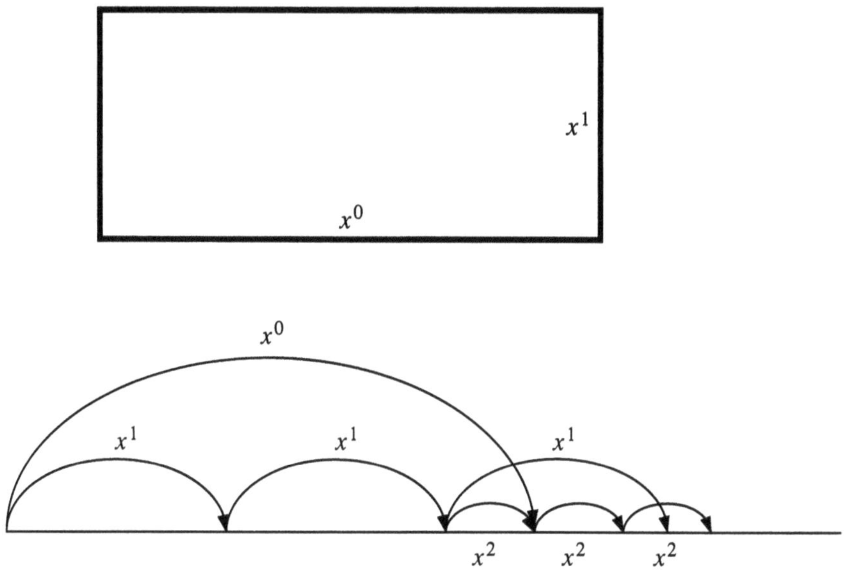

Fig. 5.11

In music we can compare the frequencies of two musical tones f_0 and f_1, and obtain a «beat frequency» $f_2 = f_0 - f_1$. Comparing f_2 with f_1, we obtain a new beat frequency f_3 — and so on. But more obvious and comfortable is to deal not with the frequency, but with the length of the wave (or length of the string). Then the situation of measurment can also be pictured by means of Fig. 5.11, where x^k is treated as the length of half a wave. The least obtained half-wave x^n plays the role of module; and the corresponding frequency $f^n = f^0 x^0 / x^n$ is the first, lowest *overtone* that coincides in two comparing musical tones.

Mathematically, the module is the greatest common divisor of two numbers. There is a well-known procedure — the Euclidean algorithm — which we can use to find this divisor.

Let us give a formal description of the process of comparison as a sequence of n procedures:

$$\begin{aligned} x^0 &= m_1 x^1 + x^2 \\ x^1 &= m_2 x^2 + x^3 \\ &\ldots\ldots\ldots\ldots \\ x^{k-1} &= m_k x^k + x^{k+1} \\ &\ldots\ldots\ldots\ldots \\ x^{n-1} &= m_n x^n + x^{n+1} \end{aligned} \tag{5.6}$$

where $x^{n+1} < \epsilon$.

The relation x^0/x^1 is approximately determined by the continuous fraction:

$$x^0/x_1 = m_1 + \cfrac{1}{m_2 + \frac{1}{m_3 + \ldots}}. \tag{5.7}$$

Let the total number of operations be constrained by the number $m = \sum m_k$. What must the optimal number of these operations (steps of subtraction) m_k be like in each procedure of measurement to provide the maximum of information?

Let the lesser segment x^k be contained in the greater one m_k times. Hence the end of the greater segment lays somewhere whithin the bounds of the $(m_k + 1)$-th step (Fig. 5.11). The «undeterminedness» of the length of the segment is thus diminished after the measurment in $(m_k + 1)$ times and the amount of information received is:

$$I_k = \log(m_k + 1). \tag{5.8}$$

The important peculiarity of this procedure of measurement is that the last $(m_k + 1)$-th step of the previous, k-th procedure is at the same time the first step of the next, $k + 1$-th procedure (the measurment of the segment x^k by subtracting the remainder x^{k+1}, Fig. 5.11). Hence the resource expense is equal to m_k (not $m_k + 1$!) operations. Now we can use the result obtained in Chapter 1. The total amount of information I, received in n procedures of measurement is equal (see (1.37)):

$$I = \sum_{k=1}^{n} I_k = I_k n = \log(m_k + 1)\frac{R}{am_k}. \tag{5.9}$$

One can see that the information is greater if m_k is smaller. But m_k can not be less than one (otherwise we do not have the procedure of measurement at all!). Therefore, the optimal, most informative value m_k is: $m_k = 1$. In this case, the sequence of the procedures (5.6) becomes:

$$\begin{aligned} x^0 &= x^1 + x^2 \\ x^1 &= x^2 + x^3 \\ &\ldots\ldots\ldots\ldots \\ x^{k-1} &= x^k + x^{k+1} \\ &\ldots\ldots\ldots\ldots \\ x^{n-1} &= x^n \qquad\qquad \text{(to within } \epsilon). \end{aligned} \tag{5.10}$$

The last (and least) segment is just the module, so that $x^{n-1} = x^n = \mu$. Let us exchange the scale of x so that the module («natural unit of measurement») would be equal to one: $x^{n-1} = x^n = \mu = 1$. Then the sequence of segments $x^0, x^1, \ldots, x^n$ forms the *Fibonacci series* taken in the inverse order. Indeed, the main property of the Fibonacci series is that the two initial items are equal to one, and each other item is a sum of the two previous items: 1,1,2,3,5,8,13,21,... The other important property of the Fibonacci series is that the relation of two consequent items of the series is equal to the «golden section» if k goes to infinity:

$$\lim_{k\to\infty} x^k/x^{k-1} = 1.618\ldots. \tag{5.11}$$

Hence, if resource R is unlimited then m can be infinite and the relation x^k/x^{k-1} can be equal to the golden section $1.618\ldots$. Otherwise (total m is constrained), x^k/x^{k-1} has to be one of the relations of the Fibonacci series. It is very interesting that the simplest of these relations corresponds to the consonances of the musical scale, namely:

1/1 = tonic,

2/1 = octave,

3/2 = fifth,

5/3 = large sixth,

8/5 = little sixth.

The other consonances can be obtained as the inversion of these ones, i.e., as a «complement to the octave». Mathematically, it is expressed as the division of 2 (= octave) by these relations:

4/3 = fourth = inversion of fifth = 2:3/2,

6/5 = little thirth = inversion of large sixth = 2:5/3,

5/4 = large thirth = inversion of little sixth = 2:8/5.

We want to emphasize that *not all* the simplest rational numerical relations are able to be consonances, but only the members of the Fibonacci series and their inversions! For instance, number 7 is not a member of the Fibonacci series and such simple numerical relations containing 7, as 7/5 or 10/7 (diminished fifth and its inversion) are terrible dissonances. Maybe it is the solution of the old damned «problem of seven», that caused such pain to theorists of musical harmony? The seven is absent among the musical consonances because it is absent in the Fibonacci series.

This model allows us also to explain the following interesting fact: we meet the golden section only in «static» arts, such as painting, architecture, sculpture, but not in music. Maybe the reason for this is that in static arts the time of observation (and, consequently, the total number of the operations m), is practically unlimited and we can arrive at the absolute maximum, which is the golden section. But in music, where the time of perception is limited *in principle*, we have to be satisfied with the conditional maximum, which gives us only the simple relations of the Fibonacci series.

Thus the main aesthetic constants — the golden proportion and the relations of the musical scale — are *optimal* from the point of view of the information maximum principle. They give us the maximum of information with the given resource expense and, hence, the maximum aesthetic pleasure. This is the reason why they are preferable in the harmonic organization of the work of art.

We enumerated, of course, only a small part of a large amount of aesthetic devices and methods, intended for the creation of harmony, «unity of variety» — on the formal, syntactical level. In the next section we will look at a more delicate example of the organization of unity in poetry — on the semantical level.

5.5 Temporal structures in poetical works: Theoretical considerations

In Chapter 3 we promised to consider one of the the aspects of the joint functioning of different mechanisms of language, together with devices that serve to make this functioning more perfect. The aspect we shall study relates mainly to the first (N1), the *associative mechanism* of language and its connections with other mechanisms, that take part in the perception of poetical texts. What is the role of this mechanism in poetry?

Usually, during the perception of poetical texts, the associative mechanism simply strengthens, supports (using its means) the images of the events (and/or objects) that the given text contains. For instance, if a winter day is described, it would be desirable for the *reader to imagine* a picture (an image) full of snow, frost, puffs of smoke from chimneys, etc. By means of the appropriate sensory associations, this (i.e., associative) mechanism assists other mechanisms, as it does also when perceiving usual (not «special») texts. Moreover, the (for poetry) specific trait of such a joint functioning of different mechanisms consists partly in their tight links during this joint functioning.

[A propos, this statement is in good agreement with Boris Eichenbaum's (1969) idea of poetry as a «game of by-side meanings of words», as well as with Vassily Nalimov's (1976) concept of the word as a «spoon scooping up meanings».]

In general, it seems possible to prove the above statement experimentally, and such topics are suited for prospective psychological investigations. But as a preliminary step to such studies, it is desirable to find some more or less simple situations, responding to the strong role of the *link between the associative mechanism and other ones*. Where can we find such «extreme» situations?

Evidently, the «assistance» of the associative mechanism is especially needed when a content of a poem is not connected (or hardly connected) with a sensual reality. Naturally, in these cases it is desirable to «colour» the text using some sensory means, in order to strengthen the entire influence of the poem. But it means that such «assistance» is mostly desirable when we deal with poetical works of an *«abstract» character* (far from a simple sensory reality), e.g., such works as rhetoric poems, didactic ones and so forth — where more or less abstract ideas are the nucleus of the poem. Texts of such character usually have no (or almost no) plot, they are more intellectual than sensual, etc. How is it possible to «colour» their abstract ideas?

Of course, there is an opportunity to «colour» a text locally, in its separate points. For instance, at a given point of the text a poet talks about the necessity of duty — and here he can compare this concept with an iron rod; then he speaks about tenderness, love, etc. — and here he can use, e.g., an image of wax; and finally, he claims that duty must be much stronger than love, and illustrates this thesis by the fact that a rod can penetrate wax. But much more often we observe quite opposite situations — when a sensory image is the primary impression of the reader, and an abstract idea is the secondary one. Such is the situation, for instance, in the poem «The Streamlet» by Thomas Hood (1799–1845):

The Streamlet

Still glides the gentle streamlet on,
With shifting current new and strange;
The water that was here is gone,
But those green shadows do not change.

Serene, or ruffled by the storm,
On present waves, as on the past,
The mirror'd grove retains its form,
The self-same trees their semblance cast.

The hue each fleeting globule wears,
That drop bequeaths it to the next,
One picture still the surface bears,
To illustrate the murmur'd text.

So, love, however time may flow,
Fresh hours pursuing those that flee,
One constant image still shall show
My tide of life is true to thee!

Evidently, here a vivid picture (of a streamlet with its waves, shadows of trees, etc.) plays the role of primary impression, and only afterwards an abstract idea arises, devoted to love and related topics (more or less abstract).

The above examples — as well as the appropriate considerations — carry no (or almost no) new concrete material concerning the empirical (i.e., poetical) reality. These almost trivial passages were presented only to accomplish the completeness of our consideration, and to show that poetical reality does not contradict our model of language mechanisms. But much more interesting conclusions about this model now follow.

The previous considerations dealt with more or less «static» situations, when no (or almost no) movement, or *development*, of a «ready» idea took place (and maybe it is not accidental that in the poem cited, Thomas Hood talked about «*one constant image*»). An idea was simply illustrated by an appropriate, more or less immobile, image or images. Quite a different situation arises when a poet wishes to deal with *dynamics*, i.e., with the development (movement) of an idea in time. In such situations, it seems quite difficult to organize the *proper sequence of sensory images*.

The main reason for this difficulty is the following. It is desirable that each work of art has a *composition* which lies not too far from the optimal one. When studying *poetical works* — because all of them are expanded in time — the most preferable composition is the so-called «*ring*»: the end of a poem has to remind one of its beginning, in order to impart the unity and structure of the poem (see, e.g., Zhirmunsky, 1975), or in order to increase the degree of its «ordering» (i.e., to decrease its entropy). The best version is a complete *overlapping* of the beginning and the end (though when the joining of these parts is not complete, but only partial — e.g., when some traits of the end do not coincide with the appropriate traits of the beginning — in such a case the information carried by these two elements is maximal). Usually such ring is formed by different means, mainly through the content of a poem, for instance, with the help of its plot. However, in the case of an abstract content (consisting of ideas), the *ring* desired can be organized only through *sensory images accompanying* these *ideas*. But what *kind of image* is suitable for this purpose? What kind of association can be used in order to provoke the appropriate sensual associations in the reader's consciousness (or subconsciousness — here we do not need to divide these concepts) while perceiving the abstract ideas of a poem?

It is clear that the *procedure* of generating these sensory associations should be an *easy* one, otherwise they cannot be born (because almost all the reader's attention is concentrated on the ideas above). This means that these images must be *close to the reader's consciousness*; in turn, such closeness responds to situations when these images are *frequently met in* the reader's *everyday life*. So we have to solve the following problem: what kind of imagery is close to our everyday life and, at the same time, can form a ring?

Our answer is very simple: those *natural phenomenae* that show a *cyclic behaviour*. Examples of such natural cycles are well known: the cycle of human life (which has its beginning «from nothing» and returns to «nothing» in the end), the seasons of a year, the 24-hour cycle, etc. All these cycles are suitable for the purposes we have mentioned above, but the last cycle seems to be the most adequate. As, in reality, the 24-hour cycle is the most frequently met periodical process; in addition, very different readers have it in common, it carries many rich images, etc. So we suspect that the 24-hour cycle (i.e., the chain: morning — daytime — evening — night — morning, and so on) would be frequently used as a basis for the *temporal* (i.e., connected with the time) *association rows* that assist some poems of abstract subject matter in organizing their imagery. How can a researcher *check* this conclusion *empirically*? One of the ways to prove this conclusion *quantitatively* is described below.

5.6 Temporal structures: Experimental results

Two experiments concerning this problem were set up, the first of them dealing with the perception of the poem «To Chaadayev» by Alexander Pushkin and the second one — with Ossip Mandelshtam's poem «Impressionism». In total, 39 subjects were involved in both experiments: 25 subjects in the first experiment, and 14 subjects in the second one. The details of the experimental procedures used, as well as some of the results obtained, were described earlier (Sheremetjeva and Petrov, 1982, 1990; Petrov, 1983, 1984/Ref. 3). Both experiments used quite identical methodology and showed quite similar results; that is why we shall present only a fragment of the results concerning Alexander Pushkin's poem «To Chaadayev» (1818). Below is the original Russian text of this poem, together with its interlinear translation into English (which we especially kept as close to the original as possible, as far as word order is concerned):

К Чаадаеву	To Chaadayev
Любви, надежды тихои славы	Love, hope, and peaceful glory
Недолго нежил нас обман,	Not long indulged us as a fraud;
Исчезли юные забавы,	Disappeared young amusements
Как сон, как утреннии туман;	Like a slumber, a morning mist.
Но в нас горит еще желанье,	But in us a desire is burning:
Под гнетом власти роковои	Under oppression of a fatal power
Нетерпеливоюдушои	Our impatient souls
Отчизбы внемлем призывање.	Heed the fatherland's calls.
Мы ждем с томлењем уповањя	We are waiting, with a languor of hope
Минуты вольности святои,	For a minute of sacred liberty.
Как ждет любовник молодои	Like a young lover is waiting
Минуты верного свидањя.	For a minute of a sure rendezvous.
Пока свободоюгорим	While we are burning with a freedom,
Пока сердца длячести живы	While our hearts are alive for honour
Мои друг, отчизне посвятим	My friend, let us devote to the fatherland
Души прекрасные порывы!	Our souls' beautiful impulses!
Товарищ, верь: взоидет она,	My companion, believe, it will rise,
Звезда пленительного счастья,	A star of a fascinating happiness,
Россиявспрянет ото сна,	Russia will awake from a slumber,
И на обломках самовластья	And on ruins of tyranny
Напишут наши имена!	They will write our names!

Here we can see that temporal (connected with 24-hour cycle) associations do really accompany the reader's movement along the text of the poem (i.e., while going from the beginning of the poem to its end), and these associations do really form a *ring*, which «models» a 24-hour cycle. This associative «temporal modelling» consists of the following. The first lines of the poem respond, in temporal associations, to night and early morning (любви, тихои, нежил, сон *love*, *peaceful*, *indulged*, *slumber*); they are followed by words that arouse «genuine» morning associations: everything around is covered by a mist, the sun rises like a giant fireball (вт̂ренннии, твман, горит — *morning*, *mist*, *burning*). Afterwards, the day comes with its vanity, matters, etc. (гнетом, власти, роковои — *oppression*, *power*, *fatal*); then we wait for the evening, rendezvous (нетерпеливою, призывање, ждем, томлењем, любовник, свидањЯ — *impatient*, *calls*, *waiting*, *languor*, *lover*, *rendezvous*); evening, sunset, and the rendezvous itself (горим, сердца, прекрасные, порывы — *burning*, *hearts*, *beautiful*, *impulses*). At last, night comes (взоидет, звезда, счастья — *rise*, *star*, *happiness*), and finally we see the morning again (вспрянет, ото сна, обломках, напишвт — *awake*, *from a slumber*, *ruins*, *write*). So we have a 24-hour *circle*, formed by the temporal associations of the words used in the poem. But still it was only a qualitative observation, i.e., a hypothesis that needed an *empirical verification*.

So the appropriate *experiment* was organized, involving 12 male subjects and 13 female ones, mainly 20- to 35-year-old graduates from different universities or high schools (several of them were students). All of them lived in Moscow and were well acquainted with the Russian literary language. Each subject had an individual conversation with the researcher.

The stimuli were the 59 words (nouns, adjectives, verbs, and adverbs) constituting the «base» of the poem. Each subject was given a set of 59 cards, on each card one of these words was printed. Nouns were presented in the nominative case, singular; adjectives were also in the nominative case and in their full form, their gender responding to the one used by the poet. This was done in order to lessen possible links between the different words which would be presented to the subjects, as well as to unify the stimuli material and lessen the probability of recognizing the poem to which the words belonged. As to the verbs, they were presented in the form used by the poet (because some forms of verbs can carry temporal meanings); the adverbs were also unchanged. Before being presented to each new subject, the cards with the stimuli were thoroughly shuffled, in order to make the sequence of the words absolutely random. And when each subject finished his answers, he/she was asked if this set of words was associated in his/her memory with a certain text. The answer was fixed in the records of the experiment.

Each subject was asked to tell what time within the 24-hour cycle was associated, in his/her opinion, with the given word. A subject could give his/her answer either in the form of an exact time or in the form of a definite interval of time (e.g., «*18–20 hours*»). In the case where a given word generated no temporal associations, this fact was designated by a dash in the records of experimental data. (The data are inclosed in brackets in cases where the time interval exceeds 12 hours; afterwards, such data were not taken into account in the processing of

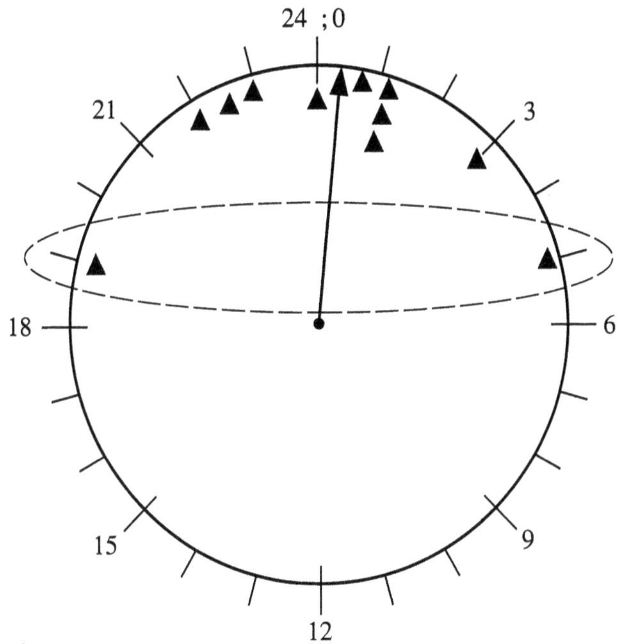

Fig. 5.12 A clock scale with experimental meanings of temporal associations of the word «ЗВЕЗДА» (*star*) in the group of 12 male subjects. Two points excluded are circled by a dotted line; an arrow responds to the mean value for the remaining data.

the experimental data.) In Table 5.1 a fragment of the records is presented concerning the 12 male subjects and the words NN40–59. (The numbers of the words presented in the first left column refer to their position in the text of the poem within a set of the above 59 words.) Temporal associations of these words have to begin from evening till night and next morning, and the experimental data are exactly of such character — as is almost immediately clear from the table.

In order to make the data obtained more suitable for further calculations, all data relating to a given word were plotted on a *24-hour clock*. Figure 5.12 presents such a clock with the data reflecting temporal associations generated by the word «звезда» (*star*) in a group of 12 male subjects. As the data of Table 5.1 show, only 11 subjects associated this word with a definite time; so 11 experimental points are plotted, each point corresponding either to the exact time indicated by a subject (e.g., 19 hours, 22 hours) or to the middle of an interval indicated (e.g., a subject N1 indicated an interval of 22–23 hours, so his point on the clock responds to 22.5 hours).

After such a procedure the primary data, relating to a given word, were analyzed for the purpose of finding a compact (more or less) *cluster*. Sometimes such a cluster is seen at once; in other cases a certain «*refining*» of the experimental data was needed. That is why in all the cases the two «extreme» points were

N	Word	1	2	3	4	5	6
40	живое alive	7-10	-	10	-	-	9
41	друг friend	18-20	18-23	18	20	-	17
42	отчизна fatherland	8-10	-	22	-	-	12
43	посвятим devote	-	-	19-20	10	-	12
44	душа soul	-	-	6	23-2	4	24
45	прекрасньыи beautiful	9-12	18-19	9	17	12	21
46	порыв impulse	18-20	15	15	20	11-12	14
47	тонарищ companion	-	10-16	10	20	-	15
48	верь believe	-	0	22	16	-	10
49	взоидет rise	7-8	4	-	2	5	5
50	звезда star	22-23	23-2	19	3	-	22
51	пленительное fascinating	-	-	23	17	6	24
52	счастье happiness	8-10	18	19	17	-	-
53	Россия Russia	7-8	19-20	6	-	21	12
54	вспрянет awake	-	-	14	17	-	9
55	сон slumber	0-9	0	5	4	23	4
56	обломок ruin	-	-	7	-	-	3
57	самовластье tyranny	-	-	15	16-20	12	12
58	напишут write	-	14-15	-	-	-	15
59	имя name	-	-	14	12	14-15	-

Table 5.1 Some of experimental results concerning 12 male subjects

7	8	9	10	11	12	Mean value	Rank
18-24	-	18-20	12	10	10-12	10.1	18
-	-	11-12	11	17	-	18.6	39
-	-	-	12	12	6	11.2	22
(8-24)	-	17-21	10	8	22-23	14.1	28
(8-24)	-	22-23	-	24	22-5	.1	51
(8-24)	-	9-10	20-21	17	-	13.4	26
(8-24)	6	-	20	22	7	17.1	31
-	-	20	20	15-16	-	18.1	33
2-4	-	-	23	24	9	.1	52
0-3	5	5-6	1	-	3	3.3	55
0-2	21-5	23	5	24	23-3	.2	53
18-24	-	-	-	17	15	20.4	46
18-4	-	-	-	-	6-7	19.2	40
18-24	-	-	16	13-14	10-11	16.2	30
18-24	7	-	22	-	22-23	19.3	41
6-8	21-8	4-5	3	(23-11)	15-17	3.2	54
18-20	-	-	12	11	-	10.0	56
15-24	-	-	10	11	11	11.8	58
-	-	-	19	23	20-21	20.8	48
(8-24)	-	17-19	10	9	-	11.2	57

excluded, i.e., those two points, without which the scattering of the remained points would be minimal (in Figure 5.12 these 2 points are circled with a dotted line). Then a mean value for the remaining data was calculated (in Figure 5.12 these 9 remaining points indicate a mean value of .2 hours).

These resulting *mean values* of temporal associations are given in the next to last column of the table. If the scattering of the remaining data concerning a given word exceeded 12 hours, we considered this word to be too «vague» in relation to temporal associations. These cases were designated by a dash in the column of mean values, but we had only 6 words «without resulting temporality» in our group of female subjects and only 1 in our group of male subjects (it was the word «святая» — *sacred*).

[It should be mentioned that the procedure of experimental data processing — dealing with excluding two extreme points and calculating the mean values for the points remaining — really joins two stages: the «refining» of the primary data, and their «aggregating» in an exact value. Their «joining» is such that the criterium of the refining (i.e., the choice of the points to be excluded) depends on the results of consequent averaging (dealing only with the remaining points). Such «joining» seems to be a useful invention, especially when it is necessary to work with a circular scale (e.g., clock), because otherwise many difficulties would arise when averaging.]

Finally, the mean values obtained were ordered — to form a *rank sequence*, i.e., a row («ordering») where the words with «earlier» associative time are followed by the words having «later» associative time. These ranks form the column to the extreme right of the table.

What were the results of this experiment?

First, it seemed reasonable to check whether the associative values (i.e., primary data given by the subjects) were *stable*. To do this, we proposed to check the associations of four words which were used twice in the poem; these four words are: сон, двша, отчизна, минвта — *slumber*, *soul*, *fatherland*, *minute*. Each of these words was presented twice to each subject, in accordance with the procedure of random choice. Associations that were generated at the first presentation of a given word were treated as relating to its first appearance in the text of the poem, while associations of the second presentation considered to relate to the second usage of this word in the text (and the experimental records respond to this convention). As a rule, the difference between associations (responding to the two presentations of the same word) was quite negligible. We attempted to use such differences to characterize each subject by an «*index of instability*»: a subject was ascribed a 1-point «penalty» when he/she showed associations without overlap and the distance between their values (or between the borders of their intervals) was more than 2 hours. (And if one of the presentations generated temporal associations, and another presentation showed no associations, a subject was ascribed half a «penalty» point.) So after the appropriate calculations, only 1 subject got a 2-point penalty, 4 subjects — 1.5 point, 6 subjects — 1.0 point,

2 subjects — .5 point, and 12 other subjects got no penalty. These results prove a very *good reproducibility* of our primary data (though a certain influence of the subject remembering his/her first associations could also contribute to these results).

As to the probable influence of the text being known to the subject as a whole — it seems to be rather small: even after receiving all 59 cards, only 4 subjects identified this set of stimuli with the poem from which these words were extracted (i.e., with the poem «To Chaadayev»).

So it seems acceptable to consider both the primary data and the resulting mean values to be quite stable.

Then the mean values were analyzed as to whether they were in agreement with our hypothesis concerning the cyclic character of temporal associations (i.e., the «modelling» of 24 hours by means of temporal associations). Such an analysis is possible by *comparing* two *rank sequences*, fragments of which are represented by the first and the last columns in the table.

The first (left) column in the table relates to the positions of different words in the text of the poem, i.e. to the *ordering of the words while reading* the text. The last (right) column relates to the *ordering of the associations* generated by these words. We supposed that the associative row should begin early in the morning, and move towards the end of the text according to the time of day. This means that these two orderings should show a high *positive correlation*. And, in fact, the Spearman coefficient of rank correlation between these two sequences appeared to approach .76 for our group of 12 male subjects (determined on the basis of 58 words of the poem which showed definite temporal associations) and .74 for our group of 13 female subjects (determined on the basis of 53 words). Besides, the associative temporal sequences for males and females were also in good accordance, showing a magnitude of the above rank coefficient near .85 (determined on the basis of 52 «overlapping» words, which had definite temporal associations both in the group of males and females); so the «internal self-consistency» of the results obtained is quite good. All three rank coefficients are statistically significant at a level higher than 99%. It means that the words within the text of this poem really have a sequence of temporal associations *coinciding* (statistically) with the movement of time; so we are dealing with the *associative modelling* of time.

Quite similar results were obtained in our experiment with another poetical text («Impressionism» by Ossip Mandelshtam) and another 14 subjects (see Petrov, 1983).

The results of these experiments allow us to draw the following conclusions:

— At least some poetical texts — mainly of didactic, rhetoric, and so forth character — possess a *structure* in their *lexical means* that generates closed associative cycles, strengthening the reader's perception of their rather abstract «content». Such a device makes the reader's perception more «vivid» through concrete sensual rows of images (maybe rooted in the reader's subconsciousness). Moreover, such a device helps the reader perceive a text as a definite integrity — concerning both its sensual and abstract content.

— From the point of view of the *creative process* such properties, when investigated, can reveal some details of the poet's style of thinking, style of life, etc. (because, e.g., Pushkin «describes» his own 24-hour cycle, typical for that period of his life, by means of the above associative row). This conclusion seems to be important also from the *gnoseological* point of view, showing that in principle it is possible to reconstruct even some «hidden», latent features of the spiritual life of the past (as well as some features of the everyday life), using a combination of a theoretical analysis with appropriate measurement procedures.

— The rather large *time distance* between the epoch of creation of the poetical works studied and the epoch of the experiment (more than 160 years for the poem by Pushkin and more than 50 years for the poem by Mandelshtam) confirms the supposition that the associative structures studied use more or less *«eternal» properties* of the human mental structures (such properties that their dependence on the epoch proves to be rather weak).

— The functioning of the associative mechanism in good accordance with our model confirms also the *general principles of the linguistic theory* proposed, which contains three mechanisms of speech perception, one of them (the associative mechanism) being rather substantial in the perception of poetical works. So, many other types of empirical investigations based on this concept and devoted to the perception of literary texts seem to be rather prospective.

*   *   *

Many various devices are inherent to art, though only some of them have been described above. It is also clear that only some of the potentially possible devices have been realized in the practice of art. Our task is not only to study the existing devices, but also to use theory to derive some new ones, and, perhaps, to assist artists in using these innovations in their creations.

Towards a human integral culture (Concluding remarks)

We have considered some problems of information theory, mainly in relation to a set of disciplines related both to science and the humanities. All our efforts were directed towards building a *general foundation* common for different branches of human activity, as if these branches have grown from a common root, i.e., the above foundation. As for the *essence* of this foundation — it is *information* and the way it functions, which seems quite natural from the *gnoseological point of view*: all objects that we can observe around us and study (using either scientific methods, means of art, or simply common sense) exist for us only because of their definite temporal *stability*. This, in turn, is caused by the fact that the *structure* of these objects is more or less stable, since the *information* (struggling with the entropy) helps to *maintain* this *structure*.

In general, the concept of information (and its maximization) appeared to be a rather adequate foundation for such various branches of human knowledge as evolution theory, physiology, psychology, linguistics, art studies, etc., and the evidence of this is that the most important laws of these spheres can be deduced from the basic informational principles. As we see the «unbridgeable gap» between science and the humanities disappearing, we notice the formation of an entire «*integral culture*» which embraces all the spheres of human spiritual life. And though we believe in science and its achievements, we also believe in the development of the humanities and art, and hope that from this side the appropriate move will be made. (Recently, it has even been suggested that in the near future — maybe around 2000 — the difference between the creative activity in art and science will vanish. Moreover, several years ago a new direction in art appeared, the so-called «constructive conceptualism», connecting art itself and the theory of art).

This new «integral culture» is now at the most early stage of its development. All kinds of activity will participate — we hope — in this process, in order to join in a new self-consistent union, science and the humanities, art and everyday life, information and creativity.

Bibliography

Akimov, Nikolaj P.

- *O Teatre (About Theatre)*. Moscow — Leningrad, 1962.

Anokhin, Pyotr K.

- *Beiträge zur allgemeinen Theorie des funktionalen Systems*, Jena: Fischer, 1978.

Ashby, W. Ross.

- *An Introduction to Cybernetics*, NY: Wiley, (1956).

Atkinson, Richard C.

- *Human Memory and the Learning Process*, Moscow: Progress, 1980.

Barrow, J.D. and Tipler, F.J.

- *The Anthropic Principle*, Oxford: Oxford University Press, 1982.

Bense, Max.

Einführung in die informationstheoretische Ästhetik. Grundlegung und Anwendung in der Texttheorie, Hamburg: Rohwolts Deutsche Enzyklopädie, 1969.

Blumenfeld, Lew A.

- *Problemy Biologicheskoj Fiziki (Problems of Biological Physics)*, Moscow: Nauka, 1974.

Breuer, R.

- *Das anthropische Prinzip*, Frankfurt am Main: Ullstein, 1984.

Broca, A. and Sulzer, D.

- La sensation lumineuse en fonction du temps, *Compt. Rend. Acad. sci.*, **134**, 1902, pp. 831–834.

Chastotny Slovar Russkogo Jazyka (Frequency Dictionary of Russian Language), Ed. by L.N. Zassorina. Moscow: Russky Jazyk, 1977.

Chekhov, Anton P.

- Sea-gull. *Sobranie Sochinenij (Collection of Works)*, Moscow: Izdatelstvo Khudojestvennoj literatury. vol. 9, 1957.

Curie, Pierre.

- *Oeuvres*, Paris, 1908.

Davies, Paul C.W.
- *The Accidental Universe*, Cambridge: Cambridge University Press, 1982.

Eichenbaum, Boris M.
1. *Lew Tolstoy: Semidesyatyje gody (Lew Tolstoy in the 1970's)*, Leningrad: Sovetsky Pissatel, 1959.
2. *O Poezii (About Poetry)*, Leningrad: Sovetsky Pissatel, 1969.

Eigen, Manfred.
- *Self-organization of Matter and the Evolution of Biological Macromolecules*, Berlin: Springer, 1951.

Fano, Robert M.
- *Transmission of Information*, New York — London: M.I.T. Press & Wiley, 1951.

Golitsyn, German A.
- Informatsija — logika — poezija (Information — logic — poetry), *Chislo i Mysl (Number and Thought)*, issue 7, Ed. by V.M. Petrov, Moscow: Znanije, 1984, pp. 9–32.

Golitsyn, German A. and Petrov, Vladimir M.
1. *Garmonija i Algebra Zhivogo (Harmony and Algebra of Living)*, Moscow: Znanije, 1990.
2. *Informatsija — Povedenije — Tvorchestvo (Information — Behaviour — Creativity)*, Moscow: Nauka, 1991.

Grigorovich, Dmitry V.
- *Dostoevsky v Vospominanijah sovremennikov (Dostoevsky in Memoirs)*, Moscow: Sovetsky Pissatel, vol. 1, 1964.

Haken, Herman.
- *Introduction to Synergetics*, Berlin: Springer Verlag, 1977.

Hoffman, Joachim.
- *Das aktive Gedächtnis*, Berlin: VEB Deutscher Verlag des Wissenschaften, 1982.

Jakobson, Roman O.
- *Novejshaja russkaja poezija. Nabrosok pervy: podstupy k Khlebnikovu (Contemporary Russian poetry. Sketch first: approach to Khlebnikov)*, Moscow: Progress, 1987.

Jakobson, Roman O. and Waugh, Linda R.
- (1987). *The Sound Shape of Language*, Assisted by Martha Taylor, Berlin — New York Amsterdam: Mouton de Gruyter, 1987.

Jaynes, E.T.

1. Information theory and statistical mechanics. I. *Phys. Rev.*, **106**, #4, 1957, pp. 620–634.
2. II. *Phys. Rev.*, **108**, #2, 1957, pp. 171–190.

Kamensky, Victor S. and Petrov, Vladimir M.

– Ob indikatorah otsenki proizvedenij iskusstva: Eksperimentalnoje issledovanije parametrov, opredelyajutshih predpochtenija. (On indicators for estimates of works of art: Experimental investigation of parameters which determine preferences), // *Sotsiologija Kultury. Problemy Sotsialnyh Pokazatelej Razvitija Kultury* (*Sociology of Culture. Problems of Social Indicators for Cultural Development*), N 108, Moscow: Institute for Culture of Russia, (1982), pp. 94–106.

Khanin, Michail A., Dorfman, Nicolaj L., Bukharov, Igor B., Levadny, Victor G.

– *Ekstremalnye Principy v Biologii i Fiziologii (Extremal Principles in Biology and Physiology)*, Moscow: Nauka, 1978.

Kharkevich, Alexander A.

– O cennosti informacii (About the information value), *Problemy Kibernetiki (Problems of Cybernetics)* Moscow: Fizmatgiz, 1962.

Khlebnikov, Velemir.

– *Tvorenija (Creations)*, Ed. by M.Ya. Polyakov, V.P. Grigorjev and A.Ye. Parnis, Moscow: Sovetsky Pissatel, 1986.

Kolmogorov, Andrej N.

– *Teorija Informatsii i Teorija Algoritmov (Information Theory and Theory of Algorithms)*, Moscow: Nauka, 1987.

Klatzki, Roberta L.

– *Human Memory: Structure and Processes*, San Francisco: Freeman and Co, 1975.

Kravkov, Sergej V.

– *Glaz i Yego Rabota (Eye and Its Work).* Moscow — Leningrad: Izdatelstvo Akademii nauk, 1950.

Kuznetsova, Ariadna I. and Yefremova, Tatjana F.

– *Slovar Morfem Russkogo Jazyka (Dictionary of Morphemes of the Russian Language)*, Moscow: Russky Jazyk, 1986.

Lanczos, Cornelius.

– *The Variational Principles of Mechanics*, Second edition, Toronto: University of Toronto Press, 1962.

Landau, Lew D. and Lifshitz, Evgeny M.

- *Mekhanika (Mechanics)*, Moscow: Fizmatgiz, 1958.

Lessing, Gotthold E.

- *Gesammelte Werke*, Berlin, 1955.

Levi-Strauss, Claude.

- *Mythologiques 1., Le cru et le cuit*, Paris, 1964.

Makhmudov, Tilab M. and Petrov, Vladimir M.

- *Voprossy Metodologii Esteticheskogo Analiza Iskusstva (Methodological Problems of Aesthetic Analysis of Art)*, Tashkent: Fan, 1984.

Martindale, Colin.

- *The Clockwork Muse. The Predictability of Artistic Change*, — Basic Books. Harper Collins Publishers, 1990.

Maslov, Sergej Yu.

- Asimmetrija poznavatelnych mechanizmov i jejo sledstvija (Asymmetry of cognitive mechanisms and its consequences), *Semiotika i Informatika: Zbornik Nauchnych Statej*, N 10, 1983, pp. 3–31.

McFarland, R.A.

- *Human Factors in Air Transport*, Design, New York, 1946.

Moles, Abraham A.

1. *Théorie de l'information et perception esthétique*, Paris: Flammarion editeur, 1958.
2. *Sociodynamique de la culture*, Paris — La Haye: Mouton, 1967.

Nalimov, Vassily V.

- *Veroyatnostnaya Model Jazyka (Probabilistic Model of Language)*, Moscow: Nauka, 1976.

Osgood, C.E., Suci, G.J. and Tannenbaum, P.H.

- *The Measurement of Meaning*, Urbana, 1957.

Panov, Evgeny N.

- *Znaki, Simvoly, Jazyki (Signs, Symbols, Languages)*, Moscow: Znanije, 1983.

Panov, Michail V.

1. *Russkaya Fonetika (Russian Phonetics)*. Moscow: Prosvetchenije, 1967.
2. *Sovremenny Russky Jazyk: Fonetika (Contemporary Russian Language: Phonetics)*, Moscow: Vysshaya Shkola, 1974.

Petrov, Vladimir M.

1. Three-block model for speech perception: an attempt of deductive construction, *Symposium on Common Aspects of Processing of Linguistic and Musical Data (Tallinn, November 22–24, 1982). Summaries*, Ed. by J. Ross. Tallinn: Academy of Sciences of the Estonian S.S.R., Division of Social Sciences, 1982, pp. 78–81.
2. Grammar as a means of association control. *Symposium on Grammars of Analysis and Synthesis and Their Representation in Computational Structures (Tallinn, November 28–30, 1983). Summaries*, Ed. by M. Tiits. Tallinn: Academy of Sciences of the Estonian S.S.R., Division of Social Sciences, 1983, pp. 73–76.
3. Evolutsija — jazyk — poezija (Evolution — language — poetry), *Chislo i Mysl (Number and Thought)*, issue 7, Ed. by V.M. Petrov, Moscow: Znanije, 1984, pp. 33–84.
4. Phoneme distinction and the sound structure of language: a deductive model for perception and information processing, *Symposium: Computational Models of Hearing and Vision (Tallinn, November 19–21, 1984. Summaries*, Ed. by M. Rauk. Tallinn: Academy of Sciences of the Estonian S.S.R., Division of Social Sciences, 1984, pp. 38–42.
5. Ob izmerenii sotsialno-ekonomicheskoj effektivnosti kulturnogo potreblenija (About measurement of the socio-economic effectiveness of cultural consumption), *«Ekonomika i Matematicheskije Metody» («Economics and Mathematical Methods»)*, **14**, N 5, 1988, pp. 832–840.
6. Sotsialno-psichologicheskije korni perestrojki (Socio-psychological roots of perestroyka). *«Slavica Gandensia»*, N 16, 1989, pp. 93–107.
7. *Prognozirovanije Khudozhestvennoj Kultury: Voprossy Metodologii i Metodiki (Forecasting Artistic Culture: Methodological and Methodical Problems)*, Moscow: Nauka, 1991.
8. Emotional impact of art: social aspect, *Art and Emotions. Proceedings of the International Symposium, September 17–21, 1991*, Ed. by L.Ya. Dorfman, D.A. Leontiev, V.M. Petrov and V.A. Sozinov. Perm: International Association of Empirical Aesthetics (USSR Branch) — The Perm State Institute of Culture, 1991, pp. 41–50.
9. Environmental features and means to survive: aesthetic signs, *Zeichen im Leben der Menschen (Signs within Human Life), For the 65th Anniversary of Prof. Martin Krampen*, Hamburg 1993.

Petrov, Vladimir M. and Yevin, Igor A.

– Emotional unity of the informational sphere: socio-psychological «climate», art and economics, *Art and Emotions. Proceedings of the International Symposium, September 17–21*, 1991, Ed. by L.Ya. Dorfman, D.A. Leontiev, V.M. Petrov and V.A. Sozinov. Perm: International Association of Empirical Aesthetics (USSR Branch) — The Perm State Institute of Culture, 1991, pp. 160–169.

Petrov, Vladimir M. and Jablonsky, Anatoly I.

– *Matematika i Sotsialnyje Protsessy: Giperbolocheskije Raspredelenija i Ih Primenenija (Mathematics and Social Processes: Hyperbolic Distributions and Their Applications)*, Moscow: Znanije, 1980.

Polak, Lew S.

– *Variacionnye Principy Mekhaniki (Variational Principles of Mechanics)*, Moscow: Fizmatgiz, 1960.

Ponomarev, Yakov A.

– *Psichologija Tvorchestva (Psychology of Creation)*, Moscow: Nauka, 1976.

Rashevsky, Nikolaj.

– *Mathematical Biophysics*, New York: Dover, 1960.

Rosen, Robert.

– *Optimality Principles in Biology*, London: Butterworth, 1967.

Shannon, Claude E. and Weaver, W.

– *The Mathematical Theory of Communication*, Urbana: University of Illinois Press, 1949.

Sheremetjewa, Natalia D. and Petrov, Vladimir M.

1. Cycles of temporal associations in poetical works: an experiment. *Symposium on Common Aspects of Processing of Linguistic and Musical Data (Tallinn, November 22–24, 1982). Summaries*, Ed. by J. Ross. Tallinn: Academy of Sciences of the Estonian S.S.R., Division of Social Sciences, 1982, pp. 88–92.
2. Temporal associations in poetical works: an experiment on perception, *XIth International Congress on Empirical Aesthetics (August 22–25, 1990, Budapest). Proceedings*, Ed. by L. Halasz. Budapest: Institute for Psychology of the Hungarian Academy of Sciences, 1990, pp. 297–300.
3. Deductive construction of the aesthetic sphere: panorama of emotion-based objects, *Art and Emotions. Proceedings of the International Symposium, September 17–21, 1991*, Ed. by L.Ya. Dorfman, D.A. Leontiev, V.M. Petrov and V.A. Sozinov. Perm: International Association of Empirical Aesthetics (USSR Branch) — The Perm State Institute of Culture, pp. 170–181.
4. La méthode statistique de comparaison des «transparences» de textes et l'éxperience de son utilisation. *Metrum of Art (Third International Conference on Aesthetics, Krakow — Przegorzaly, August 21–26, 1991)*, Krakow: Jagiellonian University, Institute of Philosophy, Department of Aesthetics, 1991, pp. 18–24.

Shevoroshkin, Vitaly V.

– *Zvukovyje Tsepi v Jazukakh Mira (Sound Chains in Languages of the World)*, Moscow: Nauka, 1969.

Simonov, Pavel V.

– *Tri Fazy v Reakcijah Organizma na Vozrastajuschij Stimul (Three Phases of Reactions of Organism on Increasing Stimulus)*, Moscow: Nauka, 1962.

Sokolov, Constantin B.

– *Sotsialnaya Effektivnost Hudozhestvennoj Kultury: Protsessy Rasprostranenija i Osvojenija Hudozhestvennyh Tsennostej (Social Effectivity of Artistic Culture: Processes of Distribution and Consumption of Artistic Values)*, Moscow: Nauka, 1990.

Stratonovich, Ruslan L.

– *Teorija Informacii (Theory of Information)*, Moscow: Sovetskoje Radio, 1975.

Suchotin, Boris V.

– *Klassifikatsija i smysl (Classification and sense)*, *Problemy Strukturnoy Lingvistiki — 1981 (Problems of Structural Linguistics — 1981)*, Ed. by V.P. Grigoryev. Moscow: Nauka, 1983, pp. 52–65.

Suna, Uldis F. and Petrov, Vladimir M.

– *Sotsiologija Esteticheskoy Kultury: Problemy Metodologii i Metodiki (Sociology of Aesthetic Culture: Methodological and Methodical Problems)*, Riga: Zinatne, 1985.

Teilhard de Shardin, Pierre.

– *Le phénomène humain*, Paris, 1959.

«Tri tsifry» iz nashey istorii («Three numbers» from our history), Argumenty i Fakty (Arguments and Facts), N 13 (494). 1990.

Trubetskoy, Nikolaj S.

– *Osnovy Fonologii (Foundations of Phonology)*, Moscow: Izdatelstvo Inostrannoj Literatury, 1960.

Vygotsky, Lew S.

– *The Psychology of Art*, Cambridge (Mass.): The M.I.T. Press, 1971.

Weber, Max.

– *Die protestantische Ethik und der Geist des Kapitalismus*, Muenchen — Hamburg, 1965.

Wilson, A.G.

– *Entropy in Urban and Regional Modelling*, London: Pion Ltd, 1970.

Woelfflin, Heinrich.
- *Renessans i Barokko (Renaissance and Baroque)*, Saint-Petersburg: Griadutchij den. 1913.

Yerkes, Robert M. and Dodson, J.D.
- The relation of strength of stimulus to rapidity of habit-formation, *Journ. Comp. and Neurol. Psychol.*, **18**, p. 459, 1908.

Zhirmunsky, Victor M.
- Kompozitsija liricheskich stichotvorenij (Composition of lyrical poems). In: Zhirmunsky V.M. *Teorija Sticha (Theory of Verse)*, Leningrad: Sovetsky Pissatel, 1975, pp. 431–536.

Zipf, George K.
- *Human Behaviour and the Principle of Least Effort*, Cambridge (Mass.): Addison-Wesley, 1949.

The Authors

German A. Golitsyn, born in 1937, graduated from the Polytechnical Institute in Perm (1959); and did his post-graduate work at the Moscow Medical Institute (1967). He received his Ph.D. in biophysics (1969). From 1967 to 1975 he worked at the Scientific Research Centre of Computer Technology; from 1975 to 1990 in the Anokhin Institute of Normal Physiology; and from 1990 to 1994 in the Moscow Institute of Radio, Electronics and Automation. Since 1994 he has been a Senior Researcher in the Russian Institute of Art Studies, Moscow, Russia.

Dr. Golitsyn is a member of the Moscow Society of Psychologists, the Moscow Society of Naturalists, the International Academy of Informatization, the International Association for Empirical Aesthetics and of other scientific organisations.

His fields of interests include the mathematical methods in biology, psychology and aesthetics, information theory, perception, motivations and emotions in behaviour, and problems of creation in science and art. He has published more than 50 works, among them (with V.M. Petrov) the monographs: Harmony and Algebra of Living (Moscow: Znaniye, 1990) and Information – Behaviour – Creation (Moscow: Nauka, 1991)

Vladimir M. Petrov was born in 1937 in Moscow; he graduated from the Faculty of Electronics, Moscow Power Institute in 1961. From 1961 to 1973 he worked at the Institute of Applied Physics; from 1973 to 1978 in the Laboratory of Sociological Models, Central Institute for Mathematical Economics; and since 1978 has served as Leading Researcher in the Russian Institute for Art Studies, Moscow. He holds a doctorate Ph.D. in physics and mathematics (1968) and also one in theory and history of culture (1993).

Dr. Petrov is Vice-President of the International Association of Empirical Aesthetics, Professor of the International Academy of Sciences (San Marino), Active Member of the International Informatization Academy, and Vice-President of the Academicís Department of Informational Culture.

His main fields of interests cover sociology and psychology of behaviour, information theory, and mathematical methods in studies of art and culture. He has published more than 250 works, among them the monographs: Mathematics and Social Processes (Moscow, 1980); Methodological Principles of Aesthetic Analysis of Art (Tashkent, 1984); Sociology of Aesthetic Culture (Riga, 1985); Forecasting Artistic Culture (Moscow, 1991).

MIX
Papier aus verantwortungsvollen Quellen
Paper from responsible sources
FSC® C105338

If you have any concerns about our products,
you can contact us on
ProductSafety@springernature.com

In case Publisher is established outside the EU,
the EU authorized representative is:
Springer Nature Customer Service Center GmbH
Europaplatz 3, 69115 Heidelberg, Germany

Printed by Libri Plureos GmbH
in Hamburg, Germany